TAINTED TAP

TAINTED TAP

FLINT'S JOURNEY FROM CRISIS TO RECOVERY

Katrinell M. Davis

THE UNIVERSITY OF NORTH CAROLINA PRESS
CHAPEL HILL

This book was published with the assistance of the John Hope Franklin Fund of the University of North Carolina Press.

Designed by April Leidig
Set in Arno Pro by Copperline Book Services, Inc.
Manufactured in the United States of America

The University of North Carolina Press has been a member of the Green Press Initiative since 2003.

Cover photo of lead pipes © iStock/Brand Diverse Solutions
Steven Barber.

Library of Congress Cataloging-in-Publication Data
Names: Davis, Katrinell, author.
Title: Tainted tap : Flint's journey from crisis to recovery /
    Katrinell M. Davis.
Description: Chapel Hill : University of North Carolina Press,
    [2021] | Includes bibliographical references and index.
Identifiers: LCCN 2020051254 | ISBN 9781469662107 (cloth) |
    ISBN 9781469663326 (paperback ; alk. paper) |
    ISBN 9781469662114 (ebook)
Subjects: LCSH: Drinking water—Contamination—Social
    aspects—Michigan—Flint. | Drinking water—Lead
    content—Michigan—Flint. | Public health administration—
    Michigan—Flint. | Water quality management—Michigan—
    Flint. | Flint (Mich.)—Politics and government. | Flint
    (Mich.)—Social conditions. | Flint (Mich.)—Economic
    conditions.
Classification: LCC RA592.M5 D38 2021 | DDC 363.6/10977437—dc23
LC record available at https://lccn.loc.gov/2020051254

For Catherine Foster White
January 18, 1940–January 24, 2014

CONTENTS

FIGURES, TABLES, AND MAPS

PREFACE

I mustered up the nerve to write about Flint, my hometown, as I drove down US 23 South toward the National Poverty Center in Ann Arbor, Michigan. The radio was tuned to an urban gospel station, WFLT 1420 AM, out of Flint. A local pastor introduced his guest, a single mother, who spoke about her trust in God despite her ongoing struggles providing for her family. Moved by her testimony, the pastor urged communities to help working parents. He cited gaps in social supports as he prayed for this mother and others in Flint facing similar hardships, and my mind drifted to my own childhood. My single mother, too, had in many ways depended on help from our church family as she raised us. How much—or how little—I wondered, had changed for Flint's working-class parents and children?

I grew up on the northwest side of Flint. We lived on Philadelphia, off of MLK Street, just a ten-minute walk from Stanley's Meat Market and its plentiful bins of penny candy and crunchy, salty, fatty, delectable junk foods. This was a working-class neighborhood, and its streets boasted bright lights, thanks to the efforts of the neighborhood block club. Things were usually calm.

By the mid-1980s, however, conditions in the neighborhood had declined significantly. Crack cocaine had arrived, just about the time General Motors plants began closing. Chevy in the Hole—the Chevrolet factory and home of Flint's 1936 sit-down strike—went in 1984, then the Buick City plant in 1987. Just before they could become members of the United Auto Workers' bargaining unit, temporary workers were hit with several rounds of pink slips.

Communities like mine began to empty out. Homeowners, including folks who accepted GM's early retirement buyouts and others frustrated with neighborhood changes, moved away. More and more, the houses weren't family homesteads but rentals. It might not have mattered before then that the neighborhood police station was closed, but now, when danger emerged, we could have used the protection. Houses were burglarized; cars were robbed in broad daylight. Crack houses proliferated, and neighborhood residents installed steel bars on their windows and bought guns for protection.

My most vivid memory of this time, as we watched the neighborhood—its identity and cohesion—fall apart, features the local grocery store Landmark. Every time we needed groceries, instead of walking the short distance to Landmark, we had to fire up my mother's mint green, boat-length Buick

Skylark and go to another store. I hated that car. It was like Christine's cousin with a busted radio. But it got us around, despite its need to stop at least every two to three blocks.

I remember asking my mother why we couldn't just get our groceries from Landmark. She'd tell me, "I ain't spending my money in there." And there were good reasons she, like the other north-end residents, chose to travel at least twenty minutes away to find another affordable neighborhood grocery store. Landmark was their store of last resort. The moment you entered, you could smell the rotting meat and stale bread. The browning vegetables limply signaled that things weren't quite right in this place. Still, the store seemed to have a consistent line of patrons, willing to pay almost double the cost of goods in other, larger Flint area grocery stores. That, in particular, struck me as odd. I was a magnet program student, touring Flint's north and south sides in my daily travels to and from school, and I knew there were grocery stores in specific communities in Flint and just outside of the city that didn't seem to be ripping people off to this degree. What made *my neighborhood* a mark for this overpriced, abysmal food? Food is a source of life. Why wasn't Landmark out of business, either forced out or abandoned by customers? Didn't we deserve the best? Like the unexplained fires that burned throughout the night and the domestic bouts that ended in bloodshed without any police intervention, Landmark was another diss to our neighborhood. A *big* diss.

So, as I listened to the mother and the pastor on WFLT, I didn't immediately think about schools or the structure of work opportunity. I thought about Flint's grocery stores, parks, and emergency services and wondered how persistent disinvestment and depopulation had impacted access to these services and amenities in Flint's poor Black spaces. I wanted to investigate the factors that undermined community access to quality essential services. Toward this end, while working on completing another book at the University of Michigan's National Poverty Center in 2012, I spent my free time pursuing these interests in Flint. I reached out to urban planners and started combing through community archives, firming up a study on the allocation of essential services in depopulating working-class cities and how these varied by race and class. Specifically, I wanted to write about how Flint and other city's residents managed the consequences of "benign neglect" as well as declines in the availability and quality of essential services.

Then, one morning, I received a phone call from a family member in Flint who mentioned that something was wrong with the water. After a summer of boil water warnings, this relative said, folks were marching in the snow claiming that the water was making them sick. When I hung up the phone, I googled these concerns; sure enough, Flint residents' complaints were vividly splashed across Facebook, YouTube, and other cyberspaces. People weren't silent about this problem. They voiced their issues with the water loud and clear, with consistency. It was the folks who were supposed to be doing the listening who were failing. Just like it had been okay to sell high-priced, low-quality food in the 1980s, it seemed it had become okay to flood the city's pipes with questionable water. Intrigued, I carefully followed Flint residents' reporting on their water issues from my computer in South Burlington, Vermont. I was happy when the semester at the University of Vermont ended, because I was free to explore what was happening on the ground in my hometown.

Days after submitting grades, I hopped on a plane headed to Flint. I touched base with locals, learning about their issues with the water. Then, back in Vermont, I processed what I had learned and began to chart out my next moves. This included sending Freedom of Information Act requests to the Michigan Department of Health and Human Services; the US Environmental Protection Agency Region 5; the City of Flint; the Michigan attorney general's office; and the City of Flint Water Department. Certain tasks were vital: I scoured piles of legal briefs and archives about Flint water. I talked to Flint residents and leaders. I reached out to Dr. Marc Edwards's Flint Water team and filed multiple Institutional Review Board requests for Genesee County private health data in order to get as close to the truth as I could. This part of the journey began in May 2015.

Since then, my work concerning Flint has been focused on expanding what we know about the factors that gave rise to the Flint water crisis as well as on the consequences of this public health disaster. A consistent finding in the sociology of disaster literature is that regulatory officials do not trust or take seriously complaints voiced by residents concerning their municipal services. I wanted to test the validity of this finding in Flint and document what happens when complaints concerning an essential service are ignored.

While writing this book, I tried to leave myself out of it. This book is about Flint and the conditions of its disinvested spaces. It's about Flint's kids and potential roadblocks to the city's rehabilitation. It's about the audacity of abandonment, the ongoing disrespect shown to Flint residents' minds, bodies, and properties. I have attempted to write this book for a general audience interested in water resource management and the politics of disinvestment, given that Flint's water crisis is an extension of a larger problem in many discarded cities: benign neglect. And so, because this book is not *just* for my academic colleagues and students of sociology, I relaxed a few standards of research reporting. You will not read hypotheses or methodological appendices. I suppressed the urge to take deep dives into empirical findings within the sociology of disaster scholarship. Instead, I have kept the focus where I believe it must be: on the people of Flint.

INTRODUCTION

The room buzzed with anticipation as people filed into Flint, Michigan's dimly lit city council chambers around five o'clock on January 13, 2014. Residents had plenty of complaints to fill their five minutes at the microphone. As they waited for the evening meeting to begin, they stamped cold feet and compared notes on the thirty-hour snowstorm that brought 17.1 inches of snow (the third-largest snowstorm here since 1921)[1]—and half the town to a standstill. It had taken far too long for the city to clear the streets in residential neighborhoods.

The residents had to brave temperatures nearing 30 degrees below zero as they worked to dig their cars out of the snow with hands and shovels, while snowplows operated by the Downtown Development Authority cleared streets in downtown Flint. The governor, Rick Snyder, promised state support,[2] but the city lacked the manpower to clear the snow from side streets.

So, one by one, Flint residents came forward. Angrily they recounted how they watched "priority areas" get plowed and how trapped they felt because of this storm and the city's response to it. They mentioned that their kids were late to school, they were late to work, and seniors were unable to go out for food and medicine. Flint citizens also offered reasonable suggestions, raised loaded questions, and boldly addressed their commentary directly to city council members, Mayor Dayne Walling, and state-appointed emergency manager Darnell Earley.

Chris Del Morone, an elderly white man, placed his notes on the podium and raised a shovel over his head. "Do we need more of these at city hall?" he demanded. Mr. Del Morone had broken the ice, and the chilled citizens crowding the council chambers broke out in laughter. "I mean, that was a lot of snow we received," he said. "It was an act of God. But, the service was terrible [and] not worth the taxes that we pay. They left the snow until the temperature got up to about 40 degrees, and it melted or rained away. That's what they did on my street by not coming until Saturday. I haven't received mail since January 4th. Thanks a lot, Flint! And then you wonder why people are leaving?" This was just one of the ways the city had neglected many of Flint's neighborhoods.[3]

Community activist Quincy Murphy, a Black man who grew up on East Marengo Street near the old Buick City plant, is known for spending a great deal of time in the city working with block clubs and churches, cutting the grass in abandoned lots, and boarding up vacant homes. He frequently voiced his concerns about the effects of neighborhood blight at city council meetings. On this day, he stood to bring attention to delays in debris removal in poor and predominantly Black areas of Flint.

"I can only speak on what I witnessed in my community where there was a big tree that fell in the backyard of one of the residential houses and tore the roof up, [damaged] the side of the house, [and] bust the windows out," Mr. Murphy said. "That situation, it still exists. The residents don't have the means or income to actually get the tree removed. Looking at the trees and the wires—I think it's an urgent situation that needs to be addressed."[4]

Murphy mentioned a resident who died after the tree fell on the utility wires and caused the power to go out in her home. "She had just got out of the hospital and was on oxygen," he recalled. She thought that the electricity would come back on, but it didn't. With his frustration mounting, Murphy complained, "I called Parks and Recreation and left messages on the phone for them and gave them the addresses. [But] it seems like it's not going to get picked up [since] we moved the tree to the side. There are several trees and branches [down] in the community. I'm quite sure it's a whole bunch of other people dealing with this same situation."

Murphy was right. Problems with diseased and dying trees weren't concentrated in his backyard or in his neighborhood. Dead trees, destroyed by parasites like the emerald ash borer and weakened by storms and old age, had become hazards. Just a few years prior, in July 2012, fallen trees trapped residents in at least ten homes on Nolen Drive, a dead-end street in Mott Park, near Kettering University and McLaren Regional Medical Center in Flint.[5]

Forestry department staffers pointed to the city's decision to lay off its workforce and to hire private contractors to address high-priority areas. All they could do, the forestry workers claimed, was direct residents to contact Consumers Energy to get on its lengthy waiting list for debris and tree removal.[6] Since dying trees continued to hang over main roads and interfere with power lines, Flint residents like Theron Wiggins have had to take on the task of removing debris if for no other reason than to clear paths on city-owned land so that mail and emergency services could reach their homes.[7]

Water bills were also on the minds of residents at this council meeting. After state-appointed emergency managers took over the city in 2010, Flint's water bills shot up.[8] Bills in Flint averaged $140 per month but just $58 in Burton, a mile up the road. Residents who were unable to pay were seeing their houses disconnected from the water system; some ended up sharing water with neighbors or pumped water from nearby wells. Increasingly concerned that they were being taken advantage of by the public utility company, people in Flint doubted the situation was going to get better.

Residents did not know why their water rates were sky-high, but they seemed to agree that the municipal problems they faced were caused by the culture of governance in the community. "It's two or three times we've come into bankruptcy. Us citizens didn't do this," said David Wilson, a Black man with a head full of graying hair who was reading from his written comments. Finding he was too fired up to read the rest, Wilson took on a preacher's thunderous tone as he pointed at city leaders. "Our politicians stole. They did it! Now they want to be bailed out by working and retired people. Every time you look around, we're in bankruptcy. I mean, [they're] just robbing us blind. We elected all you city officials, and you don't do a damn thing. Nothing! Just take from the working and retired people. What the hell we need with y'all? You can't run shit. Keep stealin'. That's all you good for is stealin' and takin' care of your buddies."[9]

It would be nearly two more years until Flint's problems with lead-contaminated water gained a national spotlight. Before news reporters, bloggers, and lawyers descended upon Flint, it was just another run-down, left-behind factory town riddled with crime and poverty. Conditions would shift when the city changed its water source in April 2014—a move that caused blood lead levels to skyrocket and residents to become sick.

One Flint resident, Jan Burgess, complained to the US Environmental Protection Agency (EPA) Region 5 on October 14, 2014:

> Earlier this year, the City of Flint, MI changed its water supply from the City of Detroit to the Flint River. This river has a very long history of pollution. Since this change, our drinking water was tripled in cost, and the quality varies daily. Some days it smells like an over-chlorinated swimming pool; other days, like pond scum. It often is brown in color and frequently has visible particles floating in it. We've been under several boil water advisories due to e-coli contamination. People in Flint have to resort to buying bottled water or having a purification system installed in their homes. Some residents have even had private wells dug. The water is not safe to drink, cook, or wash dishes with, or even give to pets. We worry every time we shower. Calls to the City and State have resulted in no action whatsoever.[10]

Burgess's complaint should have raised alarm. Instead, EPA Region 5 officials attempted to calm her concerns by insisting that water management agencies were aware of the water quality issues in Flint and were making the necessary efforts to address them. Jennifer Crooks, Michigan program manager for EPA Region 5, replied to Burgess, noting that she was copying several other top-level agency officials as well as people in the state's Department of Environmental Quality (DEQ). Crooks assured Burgess, "The M[ichigan] DEQ has been working closely with the Operator-in-Charge at the Flint Water Treatment plant to ensure a palatable drinking water is provided to citizens of Flint."[11]

By the time Burgess submitted her written complaint, other Flint residents began sharing their concerns with the local media about the odor and color changes they observed in the water flowing from their faucets. But water regulators and city leaders downplayed these problems, insisting that the water was safe to consume. Gerald Ambrose, another state-appointed emergency manager sent to Flint, called the reports isolated incidents and claimed that these issues were related to depopulation and the age of the water system.

"We understand the concerns about discoloration and odors," Ambrose said. "We tell everyone who complains that we would be more than happy to come out to their house and test their water."[12] But he didn't mention that the wait list was growing exponentially by the day or that it would take weeks for residents to get their water tested. He also didn't mention reports of waterborne disease outbreaks and verified high rates of lead contamination.

Just after the water source switch, a number of residents began to claim that the water was making them sick. Some had medication-resistant rashes and experienced stomach pains after consuming the water, while others reported more severe reactions. Gina Luster said that "she lost 67 pounds in 3 months, lost hair, collapsed at work (and subsequently lost her job due to illness); suffers from memory loss; was treated with morphine for an orange sized lump in her breast caused by 'mastitis' (bacterial infection); had a hysterectomy; and now walks with a cane." Her young daughter, Kennedy, complained of fatigue and aching bones. Meanwhile, Nakiya Wakes spotted changes in her son Jaylen's behavior by August 2014, when his temper tantrums at school were resulting in suspensions. Doris Patrick, around this time, was diagnosed with drug-resistant waterborne illnesses. And Suzzane Kolch would be the first but not the last Flint resident to die of Legionnaires' disease on July 28, 2014. (Of ninety-one subsequent diagnosed cases, a dozen would result in death.)[13]

While local leaders and water regulators stood by their claims that the bacterial issues were temporary and the water was safe for human consumption, some adjustments were made to protect state employees and the business community from the water. General Motors stopped using Flint's water around October 2014, a shutoff that cost the city almost $400,000. Later, on January 5, 2015, a week before Flint residents were told that perhaps children and the elderly should consult their doctors before drinking city water, a state building located in the city began receiving water coolers.[14]

Publicly, though, emergency managers doubled down. Early in 2015, Darnell Earley rejected the Detroit Water and Sewerage Department's offer to reconnect Flint to its previous source, free of cost. Then, when in March the city council voted 7–1 to reconnect to the Detroit department's water, Gerald Ambrose overruled the decision, characterizing it as "incomprehensible" and arguing that costs would rise despite the fact that "water from Detroit is no safer than water from Flint."[15] News reports and documents acquired through Freedom of Information Act requests later revealed the lie: instead of protecting Flint's citizens from the toxic water, state-appointed officials and water regulators actively ignored the crisis for at least a year.

Due to the inaction by water regulators and local administrators, the human toll in Flint was mounting. Victoria Marx began to notice tremors in her hands around April. Soon, she began losing her balance and having difficulties moving her right leg. She was eventually prescribed antidepressants and was diagnosed with Parkinson's. High levels of lead were found in her water, and her home's value dropped.

In October, Joyce McNeal's thirty-eight-year-old son, Joseph Pounds, was admitted to McLaren Hospital, then air-lifted to the intensive care unit in an Ann Arbor facility. By then, Pounds had "one functional lung," a body that was "ravaged by untreatable pneumonia," and a heart "surrounded by pus and [a] liver [that had] completely shut down." He died from a bacterial infection and pneumonia just two days after his thirty-ninth birthday.[16]

When the national media took notice of Flint's water crisis and descended upon the city, residents were overwhelmingly depicted as trapped victims. Outlets like the *Wall Street Journal* cast Flint residents as "the Walking Dead,"[17] while the *New York Times*, on February 4, 2016, featured the stress of anxious citizens who wished they could leave but could not afford to move away. Journalists and political pundits alike seemed comfortable with sensationalizing the pain of the Flint population, playing at determining which state or local administrator caused the water crisis and rehashing the audacious decisions that compromised Flint's confidence in its regulatory officials.

Additionally, several intriguing books emerged detailing from various vantage points Flint's struggle with its water. For instance, pediatrician Mona Hanna-Attisha offers in her 2018 book, *What the Eyes Don't See*, a firsthand account of what it took to convince people to take seriously the lead problem among children in Flint after the water source switch. Journalist Anna Clark's book of the same year, *The Poisoned City*, captivatingly chronicles the political decisions that contributed to Flint's current economic conditions as well as the steps that Flint citizens took in their fight for clean and affordable drinking water.[18]

Scholars have also written about the origins and implications of Michigan's paternalistic approach to supporting cities and towns in financial distress. Political scientist Ashley E. Nickels's 2019 book, *Power, Participation, and Protest in Flint, Michigan,* amply traces the politics of municipal takeover and details efforts made by Flint residents in recent years to restore community control. Meanwhile, political theorist Benjamin Pauli's 2019 *Flint Fights Back* is a powerful ethnographic analysis of the rise of the city's "water warriors" who shaped and fiercely protected the political narratives and goals of the anti–emergency manager movement that blossomed during the water crisis.[19]

I engage ongoing discussions about Flint in this book by relating the city's water equity issues to its systemic history of uneven development and benign neglect. Just like it took decades to bankrupt Flint's neighborhood schools, run down existing supports for the poor, and gut public institutions, including grocery stores and community centers,[20] I illustrate how Flint's water crisis was the logical result of years of service reductions, unchecked biases, and spatial inequalities during a polarizing period. In doing so, I trace the regulatory actions in Flint that undermined policies designed to protect the public from hazardous exposures, and I reveal the conditions that disproportionately affected poor and minority populations.

While acknowledging that policy choices set in motion the deliberate suppression and contamination of water resources in Flint, I argue that the delayed attempts to resolve Flint's water equity issues represent a new form of benign neglect that encourages population decline despite established regulatory protections. I capture realities of neighborhood-level neglect by detailing the journeys of Flint residents who challenged environmental injustices and the bureaucratic hurdles that got in their way. And I document the consequences of Flint's transformation from a *minimal* to a *discarded* city after the 1970s.

From Minimal to Discarded Cities: The Evolution of Public Service Provision in Postindustrial America

Just after the Second World War, thousands of Black Americans migrated to the North. They were fleeing the South, where they were restricted to menial jobs and barred from voting, but in the North, they frequently encountered job ceilings that reinforced the familiar racial boundaries and restrictive covenants that had long relegated them to the poorest neighborhoods and the worst living conditions.[21]

With the passage of the Federal Housing Act of 1949, the US government began a program of slum clearance and urban renewal that aimed to revitalize blighted areas with new housing structures and improved neighborhood conditions.[22] These efforts were unfortunately undermined by a number of factors, including Congress's failure to include an anti-segregation public housing amendment that might have mitigated the influence of racial segregation in American cities engaged in urban renewal.[23] Programs under this federal law targeted Black communities for demolition but failed to develop relocation practices and policies that would guide efforts to rehouse residents displaced by slum clearance initiatives. As a result, the widespread implementation of urban renewal or "Negro clearance" programs between 1949 and 1965 devastated many established Black communities. It was often the case that residents would be relocated to city-owned properties in worse condition than the ones that had been destroyed.[24]

The persistent push to suppress resources in poor Black communities through deliberate policy choices continued into the late 1960s. In addition to bulldozing poor minority communities and forcing former residents of public housing projects to rehouse themselves, urban planners and local officials began to embrace the policy of benign neglect or planned shrinkage that would rob these communities of the few remaining public resources that sustained them. Since overtly discriminatory policies were less politically viable in post–civil rights era America, benign neglect operated as a social engineering scheme. It was spearheaded by Daniel Patrick Moynihan, an advisor to President Nixon, and entailed cutting services in poor neighborhoods for the purposes of eliminating these "sick" communities. Moynihan suggested that the Nixon administration shift the focus from repairing and empowering to *giving* deteriorating poor minority communities *the chance to fix themselves.*[25] It was bootstrapping, neoliberal parsimony wrapped in the rhetoric of agency and community choice.

At the same time, municipalities also faced considerable economic pressures related to shifts in the global business market.[26] High-paying, low-skilled jobs became all but nonexistent in urban areas.[27] Libraries and schools closed, sometimes left vacant or devastated by arson. As a result, city-level redistributive activities were overwhelmed by efforts to remain solvent. Depopulating cities, especially those experiencing shortfalls in property tax revenues, began withholding or significantly reducing public services, including trash pickup, snow removal, and street maintenance, as well as emergency services. These developments quickly began to undermine communities' social and economic foundations.

The resulting "minimal cities," according to scholars like political scientist

Gary Miller, were primed to control costs by zoning out expensive land uses, reducing bureaucratic overhead, and developing systems of public service provision reliant upon revenue services or private goods. Higher-income residents continued to flee for the suburbs, and retail firms, light industry, and property taxes followed them out.[28] Now, such cities are discarded.

At first glance, discarded cities resemble their predecessors, minimal cities. Both share a diminished reliance on government funding as well as a commitment to eliminating redistributive city functions. However, governing styles in these cities differ substantially. Discarded cities are depopulated, resource-starved municipalities undergoing fiscal distress due to a long-term reliance on low property taxes, limited business investment, and reduced revenue support from the county or state. Demand for essential services and public infrastructure improvements, especially within low-resource, low-density, and hyper-segregated areas, aren't driven by the pleas of concerned citizens but by the impulses of the local establishment or power elite. Residents in these discarded cities are often disenfranchised by punitive policy measures, including job and housing restrictions that target felons as well as voter suppression laws designed to strip people of their political rights and opportunities to participate in the local democracy.

Within the discarded city, local government boundaries and land use policies designed decades ago to keep property taxes low and municipal services at a minimum have crippled and destabilized remaining community and public resources. Public obligations have become overshadowed by the city's urge to cut costs and privatize its capacity-building programs.[29] As a result, the gap between resources and unmet needs has expanded within these spaces. And people are facing substantial risks in these environments.

Let's imagine: While cooking dinner after work, you glance out of the window and catch a glimpse of something odd. Your neighbor is removing the siding from your home! With no delay, you call 911. If you live in a minimal city, it may take three or four hours before the police respond. If you live in a discarded city, you know from past experience that the police aren't going to show up at all. Even though you have video surveillance footage and your neighbor has a truck and a yard full of stolen siding, nothing will be done. City police are not directed to investigate property crimes. Your property is dispensable. You are expected to get over it or move away.

Under these circumstances, residents have become overwhelmed by the rising incidence of crime and blight. Back when these municipalities were on better fiscal footing and middle-class families were putting down roots, bureaucratic budgets that paid for the emergency personnel who put fires out and police officers to investigate crimes were seen as rational and necessary in efforts to provide basic public needs and services. But when cities or areas within these municipalities are discarded, governance reflects an institutional bias against the expansion of bureaucratic budgets as well as against the deployment of quality public services.

Flint, Michigan, provides an excellent vantage point from which to observe the health effects of neglect within discarded cities. To be clear: by calling Flint "discarded," I do not mean in any way to criticize or offend its residents—or those in any other city where citizens are making ends meet and trying to live their best lives under bureaucratic austerity. I refer to Flint as a discarded city because the term evokes important realities: the uneven investment, the neglected infrastructure, the silenced electorate, and the poverty that shapes and molds lives and futures every day in this space.

Once known as the "Vehicle City," Flint is a midwestern municipality that has an extensive history with discrimination and exclusionary housing practices, which relegated racial and ethnic minorities to declining areas with limited public resources. Flint was a heavily unionized and middle-class city before General Motors reduced the number of residents it employed by half by the late 1980s.[30] Like many other Rust Belt cities, Flint is now a poverty-stricken and public service–depleted space, struggling with persistent depopulation, homelessness, and crime.[31]

Traces of the city's glory years have faded away. Many spaces in Flint are now marked by overgrown weeds, littered streets, and vacated homes stripped of their aluminum siding and copper plumbing. In some neighborhoods that still have tall streetlights that burn bright at night, only a few houses remain, separated by several empty lots and boarded-up structures.

In 2017, Flint was the nation's poorest city. It ranked first in childhood poverty, with nearly 58 percent of Flint's residents under age eighteen living below the federal poverty line and a median household income $20,000 lower than the state average.[32] In addition to ranking high in unemployment and violent crime, compared with other Michigan cities, Flint ranks high in problems including infant mortality and preterm birth,[33] partially because there is so little tax revenue available to fund public health efforts.

Well-studied cities like Chicago and Detroit have seen ongoing urban renewal efforts meant to attract middle-class residents back to their city cores (and yes, those efforts are problematic in their own right). Low-income residents in discarded cities like Flint have simply had to adapt to gradual but tangible reductions in their essential services. Over the last ten years, Flint has closed more than half of its schools, the vast majority located in predominantly Black neighborhoods in the northwest part of the city. Around 57 percent of Flint's residents are Black, and yet the neighborhoods affected by school

Figure I.1. Neighborhood blight in Flint, post-2017.

closures are 90 percent Black, per the 2010 US Census.[34] Flint Northern was among four schools closed in 2013,[35] and Central High School, the oldest in the city, was closed in 2009. Officials cited declining enrollments, maintenance costs for the aging buildings, and the school system's millions in debts. Business investment and activity have declined substantially, too. Reputable grocery store chains have fled Flint, leaving its residents scrambling to find places to purchase affordable and fresh groceries.[36]

Few of the residents and community leaders I spoke with were shocked by this public service exodus. At the same time, many contend that Flint's "criminal element" has made a bad situation much worse. As one resident put it, "I've passed two blocks of houses. No windows, no siding, no doors. Two blocks. They stole everything off these houses. You get down to the end, there is one house on that side and another house across the street. Hell, the neighborhood is fucked. That's why Meijer's [grocery store] closed. They've been robbing people at Christmas and selling DVDs. Carjacks and kidnapping [were happening] right there on Pierson Road. They were hanging out in the parking lot with their trunks open. I stopped going to that store years ago."[37]

It's not that city officials are uncaring or willfully resistant to improving conditions; they simply cannot meet even basic service needs with the available resources. At this point, discarded cities are so broke that, if they could fund projects, circumstances likely would still prevent them from completing the work. Their property taxes have, like their residents, moved to the suburbs (almost 90 percent of the tax base in cities like Detroit is concentrated in suburbs).[38]

Even there, residents have been hit hard by precarious job conditions, especially those living in racially diverse working-class and "industrial" suburbs, where concentrated poverty has skyrocketed.[39] As a result, despite the suburbanization of low-wage retail sector jobs, the mismatch between affordable housing and lucrative occupational opportunities in these spaces has widened. This is particularly the case in areas like Burton in Genesee County[40] that are still reeling from widespread layoffs and factory closings.[41] Families who initially moved to the suburbs to flee the crime and the blight of urban areas increasingly face the same problems they attempted to avoid.

Why State Actions Matter: Michigan's Role in Bankrupting Cities and Its Attempts to Take Them Over

Without a big push from the State of Michigan, Flint would have had a tough time transforming into a discarded city. In its latest round of disenfranchising Michigan cities around 2011, state lawmakers embraced a governing strategy that enabled the state to snatch the reins by bankrupting struggling

Figure I.2. Neighborhood upkeep in northwest Flint.

municipalities and replacing their elected leaders with appointed administrators serving at the governor's pleasure.

Impact of Changes in Statutory Revenue-Sharing Formulas

Michigan began this process by reducing the funds available for the purposes of delivering public services. Rather than implement capacity-building strategies and initiatives designed to invest in families and help sustain the social contract between employers and the working class, the state altered its revenue-sharing distribution procedures. Thus, as layoffs multiplied and low-paying part-time jobs flooded the labor market, the state gradually bankrupted nearly 500 cities.[42]

Statutory revenue sharing is, essentially, the legally mandated distribution of a pot of money set aside by the state constitution to cover essential city services such as fire services, road repair, water and sewer service, garbage pickup, and police protection. Although statutory revenue sharing is not guaranteed year to year, Michigan's cities and counties had become reliant on these funds as other economic effects took hold (again, including eroding property tax bases and the decline of manufacturing jobs). However, despite the consistent growth in Michigan's state tax revenue over the past decade, the state legislature's formula for establishing annual state revenue-sharing payouts to municipalities has become less lucrative.

In 2001, $900 million from statutory revenue sharing was split among Michigan's 485 municipalities. Via Executive Order 2002–22, then governor John Engler drastically cut this revenue sharing to $847 million, altering the distribution formula. By 2011, this number had plummeted to $215 million when Rick Snyder, a lawyer, accountant, and self-proclaimed "tough nerd" from Battle Creek, Michigan, was elected the next governor and the local funds from statutory revenue sharing were reduced by another 30 percent.[43] Since then, only around $30 to $40 million of those funds have been distributed to municipalities annually.[44] As a result, between 2003 and 2013, Detroit lost $732 million and Flint lost almost $55 million as the state diverted funds traditionally used to maintain city services to state budget needs. Cash-strapped Utica has around 4,700 residents and lost about $1.4 million between 2003 and 2013; its mayor, Jacqueline Noonan, proclaimed, "It's like somebody stealing your wallet and then coming back hours later and saying, 'What, you have no money?' . . . It's ridiculous. It's insane."[45]

Impact of Emergency Manager Laws

While extracting funds from cities and undermining efforts to make *all* Michigan communities stable and livable places, state-level Michigan officials put in motion a set of administrative rules that extended their capacity to govern jurisdictions deemed to be "in financial distress." Essentially, the state gave itself the right to ignore elected officials (and the voters who put them in office) and install state-appointed administrators. In 1988, Public Act 101 granted the state the right to appoint an emergency financial manager to assist cities in financial distress, and in 1990 Public Act 72 was passed, allowing the governor to declare such a financial emergency (though elected officials were not to be removed and the emergency financial managers' authority extended only to financial matters).

The emergency managers could renegotiate terms with service providers but could not break union contracts. The first emergency managers were appointed by Republican governor John Engler in 2000 and 2002 and assigned to avoid bankruptcy filings by the cities of Hamtramck, Highland Park, and Flint. Though their efforts did not achieve long-term success, emergency managers were soon sent to take over school districts in poor Black communities including Highland Park, Ecorse, Benton Harbor, and Pontiac, as well as Detroit's public schools.[46]

There has been broad, persistent bipartisan support for emergency managers in Michigan, though they have a history of abusing their powers. For instance, former Pontiac emergency manager Michael Stampfler faced fierce criticism when he chose to contract Pontiac's water treatment to United Water, a company facing twenty-six felony indictments for violations of the Clean Water Act in 2011 after manipulating wastewater sampling methods and disinfectant treatment levels before sampling periods in Gary, Indiana.[47] In another example, Arthur B. Blackwell II, a Highland Park emergency financial manager appointed by Governor Jennifer Granholm in 2005, was ordered to repay $264,000 in restitution and serve two years' probation for fiscal malfeasance—essentially, paying himself improperly.[48]

Many Michiganders opposed the state's efforts to empower emergency managers, who assumed the duties of elected officials. Michigan voters demonstrated their opposition by getting this issue on the ballot and repealing Public Act 4 in the November 2012 general election. However, the Michigan legislature and the Snyder administration soon showed that their shared drive to diminish the authority of local officials in poor cities and school districts would not be deterred. Senate Bill 4214, which developed into Public Act 436, was introduced on December 26, 2012, a mere forty-nine days after Michigan voters rejected Public Act 4.[49]

Like Public Act 4, Public Act 436 affirmed the emergency managers' right to "rule by decree" over municipalities in financial crisis. They could "exercise any power or authority of any officer, employee, department, board, commission or other similar entity of the local government whether elected or appointed."[50] These powers are predicated on a major assumption: that local leaders are incapable of resolving financial crises or making the hard but needed cost-cutting decisions that can lead to fiscal stability.

During Governor Snyder's administration, communities under emergency management doubled. Notably, the move to appoint an emergency manager has been disproportionately utilized in poor Black cities. By 2017, only 3 percent of Michigan's white population had been under an emergency manager's control, but more than half of Michigan's Black population was governed by unelected and state-appointed officials.[51] As the state's citizens complained about the hidden process of declaring a city "in fiscal distress," the state government developed a Fiscal Health Score. This supposedly transparent and impartial measure consisted of a ten-point scale: jurisdictions with scores of 0–4 were considered "Fiscally Neutral," cities with scores of 5–7 were put on the "Watch List," and cities scoring 8–10 were considered to be in fiscal distress.[52]

Still, emergency managers were frequently deployed to Black communities while predominantly white cities with higher fiscal scores remained governed by their elected officials. For instance, an emergency manager was sent to Pontiac (52 percent Black American) but not to other Oakland County cities with the same Fiscal Health Score, including Troy (only 4 percent Black American population) and Hazel Park (10 percent Black American). In Genesee County, Genesee Township (90 percent white) received a score of 9 and Flint (predominantly Black American) scored an 8. Only the latter was selected by the state treasurer to receive an emergency manager.[53]

The State of Michigan began interrogating Flint's finances with a review by Andy Dillon, the state treasurer, in August 2011 under Public Act 4. Flint's 2011 audit revealed a $25 million deficit, with about $9 million attributed to the water supply fund. Governor Snyder then assembled an eight-member team, which recommended, on November 8, 2011, appointing an emergency manager in Flint.

Snyder tapped Flint's former interim mayor Michael Brown on November 29. Brown's directive in Flint was to "simply do one thing and one thing only, and that's cut the budget—at any cost."[54] Not even seventy-two hours later, Brown had eliminated the salaries for Flint's mayor and city council. A week after that, the offices of the ombudsman and Civil Service Commission were closed. Then, seemingly overnight, water rates in Flint began to skyrocket.[55]

This chain of events shows that Michigan's method of rescuing discarded cities from bankruptcy and struggling school districts from closures has led directly to serious sociopolitical and health ramifications for its residents. And yet the emergency manager law survived a court challenge. The fact is,

Michigan's brand of neoliberalism—which includes disempowering elected city officials and bringing disinvested communities to their knees in the name of cost cutting—has received far too little scrutiny, then and still today.

Assessing Flint's Road to Recovery and Remediation: The Roadmap

Contemporary efforts to suppress vital resources stem from the long-standing policy of benign neglect in poor and minority spaces. Communities within Flint, undermined by years of disinvestment and apathy, continue to be manipulated by a confluence of local conditions that encourage dysfunction, disease, and despair. For decades, social scientists have been interested in how groups work together to motivate social change within poor and discarded cities like Flint.

Numerous studies have documented the limits and rewards of collective efficacy, which consists of a group's shared belief and execution of a course of action within community settings.[56] Most of this scholarship in recent years draws from the generative work of sociologist Robert J. Sampson and his colleagues, who used collective efficacy theory to understand how Chicago residents responded to crime in their communities. With Sampson's influence, scholars have focused on relating collective efficacy to social cohesion, defined as mutual trust and solidarity among unrelated community members, and informal social control, a concept that refers to civilian actions taken to intervene in community problems.[57] The literature in this vein generally agrees that the communities that accomplish the most through their collective action have figured out how to come together despite their differences to accomplish a common goal. In these studies, social networks among community residents are a prerequisite for sustained and effective community action.[58] Examples include working-class communities conquering health emergencies by pooling resources from neighborhood institutions like Masonic lodges, churches, and block clubs to help vulnerable residents gain access to vital resources during crises.[59]

A few scholars have also commented on how local governments contribute to "poverty management,"[60] detailing the ways states govern cities in financial distress.[61] Still, little is known about the consequences of decades of neglect and how neoliberal governing regimes relate to downward shifts in health and quality-of-life outcomes among residents.[62] We know that many municipalities are essential service providers, and so it is crucial that we examine the relationship between public health and the quality of municipal services. Cities have restructured debts and stripped down services in order to balance commitments to retired public employees, creditors, and current residents. And while cities must protect public safety, even when subject to bankruptcy and receivership laws, statutes are deliberately vague about what constitutes "safety" and which services are essential.[63]

My efforts to capture the health effects of disinvestment and neglect in an era of "personal responsibility" were made possible by people who welcomed me into their homes and offices and made time in their busy schedules to speak with me. Given my intent to capture how community leaders and local-level administrators viewed the water crisis in particular and how their standpoints shaped their responses to it, I made efforts to reach out to people who expressed their commitment to community empowerment, online or by deed. Toward this end, I scoured the Internet for neighborhood groups and individuals in Flint working to improve community conditions and services. I made attempts to reach out to people in Flint who had been vocal about the water problems, including community group members, church pastors, and those involved with nonprofit agencies. I contacted these community leaders and advocacy groups to get their take on the city's current water issues and to learn about their approach in advocating for the resolution of these issues as well as to hear about ongoing concerns with essential service gaps in Flint.

At this time, many of the residents who would eventually become vocal critics of the crisis had not emerged. My initial attempts to find community leaders to speak with were not particularly successful. Several community organizations disbanded. Others never responded to my attempts to engage. I was bounced around from contact to contact until I was able to secure interviews with a small sample of leaders with diverse opinions in the community.

While I conversed with a few other councilpersons, I spent the most time with First Ward city councilman Eric Mays, a middle-aged Black American and proud Michigan State University graduate. He was generous with his insights on the crisis as well as on the economic concerns that Flint must face to survive this ever-changing global economy and political landscape. Councilman Mays has been seen on television, angrily contesting being tossed out of meetings by county sheriffs and wearing handcuffs as he attended a court hearing for impaired driving.[64] At the same time, those who regularly attend city council meetings have also become accustomed to listening to Mays's booming, deep voice and witnessing his passion for Flint and the people he serves in the First Ward, where I grew up, on the northwest side of the city.

The insights in this book were also shaped by interviews with unelected residents working to make a difference in Flint. Our conversations occurred before the problem with lead contamination was exposed to the public, and so our exchanges unearthed contrasting views about the crisis. For instance,

Bill Hammond, a retired white community leader with Flint Neighborhoods United, who graciously spoke with me about the water equity concerns in Flint that had started making news, confidently claimed that complaints with the water weren't pressing. Instead, Mr. Hammond was cautiously optimistic about Flint's economic future and the health of its depopulating and poverty-stricken neighborhoods.

On the other hand, community leaders like Melissa Mays (no relation to Councilman Mays), a white working-class married mother of two who had recently moved to Flint, completely contradicted Mr. Hammond's stance on the water as well as his trust in the competence of local and state leaders. Mrs. Mays vividly described how the water quality changed in late April 2014 and the extent to which this issue made her family sick. Despite holding down a full-time job, Mrs. Mays responded to the urgent need by seeking out the source of their sickness, then going out to plan water equity marches, engage area churches for support, enlist outside guidance from environmental activists like Erin Brockovich, and share information about the water through her website.

Mrs. Mays gave me no impression that she expected Flint leaders to fix this problem or any other pressing issue facing the city. She believed that the people whom citizens trusted to lead the city failed them, offering nothing but defensive deflections and misleading assurances. In my three-hour interview with Mrs. Mays and in my brief conversations with other Flint residents during this time, I saw how the water crisis had raised suspicions. Maybe, residents thought, it was just another state attempt to deliberately destroy and perhaps replace the poor and working class in this community.

My effort to examine the health effects of neglected space and how resident leaders have responded to these issues also involved developing an understanding of Flint's history of distributing resources and engaging with citizen complaints about the neglected or abandoned essential services. I undertook extensive archival research in the University of Michigan–Flint's Genesee Historical Collections Center, the Bentley Historical Library at the University of Michigan in Ann Arbor, and *Flint Journal* archives housed at the Flint Public Library. This data helped me create a social history of water quality concerns and issues with essential services in Flint from the mid- to late 1960s to just before the water source shift in 2014.

In September 2015, Melissa Mays contacted me to let me know that civil engineer Marc Edwards had been tapped to help get the water tested. I was excited—this seemed like a step in the right direction. Dr. Edwards's team validated what residents already knew: their water was poisoning them.

I reached out to his team and requested access to their data in hopes of identifying patterns of water neglect in the community. I was trying to gain a sense of where the water problems were located across neighborhoods and to develop a social demographic profile of citizens disproportionately impacted by the city's water system neglect. Dr. Edwards graciously shared the water testing data—but I didn't see any systemic clustering of lead contamination in his team's tests of Flint homes. While there were plenty of homes with extremely high levels of lead, most of the homes in the sample came from neighborhoods with similar characteristics.[65]

Officials at this time were whitewashing the situation and giving residents every indication that their problems with the water were imaginary—or, if real problems existed, they suggested it was the homeowner's job to fix them. The discovery by Dr. Edward's team was a big win for Flint. Residents finally found proof that the water was contaminated.

I was determined to generate a profile of the residents impacted by water neglect, so I went through a frustrating but enlightening Freedom of Information Act stage in my research. Any state office I thought might have data on Flint's water received a public information request. My requests were about everything from shifts in Flint's water chemistry to water analyses and inspection reports, audits detailing water system maintenance issues, and documented resident complaints. US Environmental Protection Agency Region 5 received a request for any business or consumer complaint letters, emails and phone logs pertaining to Flint's water issues through the Safe Drinking Water Act Hotline between 2010 and 2015, and complaint letters and emails from consumers and businesses forwarded to the Water Resources Division and the Pollution Emergency Alerting System.

It would take months of back-and-forth, concessions and delays, before I was surrounded by over 50,000 pages of documents. The process of combing through these documents was tedious, but my review of the EPA information made it clear: I had personally spoken with more people in diners and coffee shops and listened to more resident leaders at various city council meetings discuss their water problems than were represented in the forty complaints that the state and the EPA turned over to me. I knew that this information had huge holes. So, I set out to fill them, religiously inquiring about—and cajoling replies to—my FOIA requests.

This effort led to a brief but pivotal phone exchange with Rita Bair, section chief of EPA Region 5's Ground Water and Drinking Water Division. I asked Bair why the file of water complaints made by Flint residents was so thin. She told me that they did not have any mechanism for recording or logging the calls placed with the Safe Drinking Water Act Hotline. So, when people called, I asked incredulously, her department essentially dropped their complaints into a bottomless pit? Water regulators *had* to have known that nothing would ever come out of these complaints, because no one was positioned or even capable of retrieving them to effectively address the issues shared by residents.

Outrageous. This was yet another diss to the community and a huge waste of people's time. Who in Flint had time to make a complaint, about their water no less, and then make it over and over again? Most "Flintstones" are poor people who work all day, and sometimes for multiple employers, just to stay broke.

I continued to stew over the missing complaint data until another idea came to me. Instead of attempting to learn about the distribution of the water problem throughout Flint's community by way of complaints, I would follow this trend by tracking the presence of lead in children's blood in Flint before and after the April 2014 water source switch. I also wanted to trace the potential correlates of lead by relating the distribution of lead exposure in the community to the spread of neighborhood neglect.

It took almost a year and a half to get this data. Months before my data request, Dr. Mona Hanna-Attisha, pediatrician at Flint's Hurley Medical Center, ran into the same brick wall when attempting to analyze blood lead trends among preschool children in Flint. Thankfully, Dr. Hanna-Attisha had the juice to gain access to the data administratively before securing permission from an institutional review board to pursue the study, thereby avoiding a long wait and enabling a comparison of blood lead levels, both pre– and post–water source switch. Dr. Hanna-Attisha's findings were another win for Flint residents: her study proved that Flint's lead exposure problem had also affected the health of its children.

Once my study was approved in 2016, I received (after months of waiting) the private health data for Genesee County children aged eleven and under from the Michigan Department of Health and Human Services (MDHHS). Since 1997, all laboratories in Michigan have been required to report blood lead screening results to the MDHHS. Blood lead level results, maintained by Michigan's Childhood Lead Poisoning Prevention Program, provide information about each individual screened, including his or her address, race, sex, birth date, method of sampling, blood lead screening result, and date the specimen was collected.

I used this private health data from the MDHHS to describe the demographic characteristics of neighborhoods heavily burdened by water quality issues and to determine the extent to which these issues were correlated with additional forms of neighborhood disorder (for example, brownfield sites, homes built before 1950, and water main breaks). I also used this data to document the extent to which children with lead poisoning received follow-up testing before and throughout the water crisis.[66] I accomplished these tasks by linking the data provided by the MDHHS to US Census data as well as to enforcement data from the EPA, water main break data from the City of Flint, water testing data from the State of Michigan, and data on brownfields from Michigan's Inventory of Facilities, reported as authorized by the 1994 Natural Resources and Environmental Protection Act, Public Act 451.

Finally, in light of persistent speculation and criticisms surrounding Flint's water recovery process, I made my last FOIA request to the City of Flint. Although the city launched its Flint Action and Sustainability Team (FAST) Start pipe replacement program in March 2016, Flint residents continued to complain about water issues.[67] Throughout the first five phases of the pipe replacement project, the city's progress was inconsistent, especially throughout Phases 1–4 (May 2016–April 2018). In my FOIA request (19–0182), I requested access to address information (street address and zip code) for all service lines that were examined and repaired to date as part of the service line replacement process.

After learning that my FOIA request was approved, I was dismayed once again to find out from the city's FOIA coordinator that the water department's accounting of this information was accessible only through the city's FAST Start initiative website. The site lists the following information for each property (just under 3,100 homes with designated repairs) on a map interface that is sortable by phase (1–3 and 4): house number, street name, parcel ID, and replacement type. I extracted all records from Phases 1–4 from this website and cross-referenced this information with Genesee County parcel data and 2017 neighborhood data at the census tract level for Flint in order to account for socioeconomic and demographic differences between the areas that received repairs and those that were overlooked.[68]

Organization of the Book

Part 1 describes conditions in Flint before residents' water source concerns were validated by Marc Edwards and his team in September 2015. Chapter 1 sets the stage of the study by documenting the fate of efforts during the Model Cities program era to reverse trends of neighborhood neglect in Flint and position citizens to play a more significant role in the city's governance. In this chapter, I also address how persistent race and class discrimination shaped the distribution of public services within Flint, as well as how these factors influenced neighborhood conditions available to residents.

In chapter 2, I examine the legacy of efforts to improve the community relations and public infrastructure in Flint. The evidence I present in this chapter documents the extent to which decades of abandonment have affected the quality of essential services available to Flint citizens. Data from archives, news reports, and institutional documents illustrating the mounting environmental hazards and the deteriorating conditions in Flint between 1985 and 2013 are featured in this chapter. Chapters 3 and 4 document the drinking water concerns voiced by Flint residents and how community leaders approached the water crisis before it became national news. In these chapters, I also detail residents' struggles to address environmental hazards and gaps in essential services and how neoliberal actions by the state undermined citizens' efforts to intervene and change the effects of benign neglect.

Part 2 of this book documents attempts to address the water crisis in Flint after concerns with the water were backed by independent research. Relying on internal EPA, DEQ, and City of Flint memos; lawsuits; petitions; FOIA requests from the EPA; and news reports to illustrate the structural factors and institutional actors involved in undermining the efforts of community groups, I document in chapter 5 how misinformation and apathy affected the City of Flint and DEQ officials' capacity to govern and regulate the public water system. I also examine how the resilience of community action was preserved and supported in the face of structural and political hurdles by academic research, interracial alliances, and persistent advocacy. In chapter 6, I document the lawsuits that emerged in response to the crisis, in addition to the conflicts that developed among key players in the water crisis due to opposing views on how best to move forward after evidence of the water contamination surfaced.

In chapter 7, maps from the geospatial analysis, which relate lead contamination and blight, help illustrate the distribution of neglect as measured by lead-contaminated drinking water results, blood lead levels among preschool children, and neighborhood blight. In this chapter, I also document the initial attempts (Phases 1–4) to repair Flint's lead service lines and how those efforts to get the lead out of Flint's water were sidetracked by various conflicts between the City of Flint and the State of Michigan as well as between the Flint City Council and the company hired to oversee Flint's FAST pipeline replacement program, multinational engineering firm AECOM. Although Flint possessed funds to proceed with repairs, I illustrate in this chapter the extent to which disagreements regarding how best to proceed, even concerning locating the lead service lines that needed repairs, undermined Flint's pipe replacement efforts. I conclude this study by reflecting on how structural constraints and actions by regulatory officials shaped the rewards of collective efficacy or, more specifically, the Flint community's tenacious and strategic activism throughout the water crisis. I address how water quality issues aren't unique to Flint and why citizens in other cities in the United States should be curious if not cautious about their drinking water. And I comment on the status and implications of Flint's recovery from this devastating—and unresolved—public health disaster.

This book is a cautionary tale about what happens when American cities are discarded and citizens are left behind. Understanding how systematic efforts to disenfranchise working-class and minority communities in Flint impacted health outcomes helps bring into focus the often overlooked and underexamined consequences of America's prolonged war, not against poverty, but against *poor people*. To be sure, community opposition to uneven development, persistent racial inequality, and benign neglect has been fierce and indomitable. Every step of the way, the people of Flint have fought for dignity and accountability. They continue that fight today and will likely continue to engage for years to come.

PART I

Before Evidence of Lead
Contamination Surfaced
in Flint's Water

1

Flint during the Model Cities Era

The Challenges to Community Advocacy

By the early 1960s—after picketing, staging boycotts, and enduring white mob and police violence—little had changed in the lives of working-class Black Americans. Frustration was building.[1]

In northern states, attempts by Black Americans to secure suitable housing were quashed. For example, civil rights groups in Chicago initially led by Martin Luther King Jr. formed the Chicago Freedom Movement. Participants declared a war against decaying and segregated slums—where arrests were high and services neglected—with rent strikes and marches to protest slum conditions. But their efforts were undermined by city officials who refused to work with citizens or the civil rights establishment to address the lack of public services in poor and segregated communities. Although Black Americans and other poor racial and ethnic minorities typically paid a higher percentage of their monthly earnings for housing than did whites, their neighborhoods were run-down, with deteriorating homes and poorly maintained tenement buildings. This was especially true in cities like Flint, Michigan,[2] where Black Americans were relegated to segregated housing rife with crumbling infrastructure (while they scrounged for work in an increasingly dead-end labor market).[3]

Despite the passage of public laws like the Federal Housing Act of 1949—implemented to provide suitable homes and living environments for all Americans—major cities in Michigan and throughout the United States continued to be marred by substandard and blighted housing stock.[4] Local leaders and planners undermined actual living conditions by embracing destructive urban renewal and community development projects that sought to lure businesses and the middle class back into cities, while municipalities further weakened communities by opting to overlook—rather than intervene against—the rigid patterns of housing segregation, which included a dual housing market that relegated Black Americans to the parts of cities most hampered by substandard accommodations. This reality typically contributed to disproportionate rates of health issues in Black communities, including tuberculosis, asthma, measles, and low-weight births.[5]

In cities like Detroit, where the automobile industry brought not only employment but also funding for urban renewal programs that financed highways and skyscrapers, Black neighborhoods were being dismantled. Federal urban renewal programs took jobs and resources to white suburbs by routing highways, such as Detroit's I-375 and Flint's I-475, through Black neighborhoods— wiping them out and displacing their residents.[6]

The Model Cities program, authorized by the Demonstration Cities and Metropolitan Development Act of 1966, was a federally funded initiative implemented to improve poverty-stricken areas within selected cities. It was originally conceived in October 1965 by the Task Force on Urban Problems appointed by President Lyndon Johnson and chaired by Robert C. Wood, a professor of political science at the Massachusetts Institute of Technology. After some consideration, the task force generated a report recommending that President Johnson develop a five-year program that targeted and coordinated existing government resources to address blight and disenfranchisement in the nation's largest cities. Embraced by the Johnson administration, this program inspired the passage of the Demonstration Cities and Metropolitan Development Act and was considered a solution to urban violence and a response to frustration with Great Society programs. Congress allocated $924 million to jump-start this initiative and granted the US Department of Housing and Urban Development (HUD) the authority to develop a selection process that would determine which cities received funding. With the help of the Washington Interagency Review Committee—consisting of representatives from the Office of Economic Opportunity as well as the Departments of Labor and Health, Education, and Welfare—HUD announced that sixty-three cities were initially selected. Later, in March 1967, another twelve cities were included, followed by seventy-two more in the autumn of 1968.[7]

In many ways, Model Cities was a unique policy intervention. It primarily attempted to intervene in resource-deprived areas by revisiting how public services were provided and uniting citizens and local politicians in plans to address local problems. Under the watchful eyes of Regional Interagency Coordinating Committees meant to monitor local activities, Model Cities provided opportunities for struggling municipalities to revitalize neighborhood conditions, rejuvenate community relations, and improve the range of employment and housing options available to residents of blighted spaces.[8] This program also made attempts to address interrelated social problems, ranging from inadequate housing and underemployment to disparate health outcomes and physical disorder.[9]

It turns out that race- and class-based antagonisms diluted the program's impact well before Congress withdrew funding in 1974. Although citizen participation was required in almost every phase of decision making,[10] efforts to dismantle long-standing structural exclusions and inequalities with Model Cities initiatives in Flint and other US cities were undermined by a consistent conflict between citizens and local leaders. Model Cities efforts also struggled to live up to the program's potential because these initiatives could not overcome or neutralize entrenched opposition from local elites who challenged attempts to desegregate public spaces and improve the quality of public services and accommodations for all residents. Like Flint, many of the communities that implemented Model Cities plans realized they needed a more substantial and long-lasting financial investment. A gamut of issues needed to be addressed, including the need for better jobs, higher quality and affordable housing, better schools, and an end to police abuse. Model Cities did not ultimately meet the needs of these communities.

I document in this chapter how, after the Model Cities program concluded, many of the problems that it was designed to alleviate blossomed: troubled schools, high crime rates, and deteriorated housing. Further, the shift to neoliberal policy making on the national level and lagging faith in the effectiveness of public participation in poor communities severely undermined efforts to address neighborhood blight through collective action effectively. In Flint, this set the stage for a public health crisis to explode decades later.

The Model Cities Program: Hope for a Greater Society

Before Model Cities initiatives were implemented in Flint, Black Americans faced nearly insurmountable barriers to improving their segregated neighborhoods or escaping these dilapidated spaces. White southerners—overwhelmingly from Arkansas, Missouri, Kentucky, and Tennessee—had migrated north for General Motors jobs and made up a heavy presence. Lingering Jim Crow social dynamics legitimized the harassment and humiliation of Blacks in American society and were amplified in Flint.[11]

With few exceptions, pre–World War II Black Americans were denied service in many of the city's public establishments, including restaurants, hotels, bowling alleys, skating rinks, and taverns. They were confined to the neighborhoods south of downtown near Thread Lake. City officials, business elites, and the real estate apparatus helped to protect segregationist norms, especially within Flint's schools, workplaces, and neighborhoods. Police officers protected strict color lines, including enforcing the Flint River demarcation that separated St. John Street from the city's predominantly white east side. Black Americans were prohibited from attending school or being found "walking up and down the street" on the wrong side of this line unless they were employed by a white person in the area.[12]

In the 1940s, the white St. John Street community was dominated by immigrants from southern and eastern Europe. Over time, it would emerge as a business epicenter for Black residents, whose share of the Flint population increased by more than 400 percent between 1940 and 1960 (from 6,559 to 34,521) as black southern migrants flocked to Flint's factories.[13] As Blacks moved into this area, white families and small businesses moved out, creating hardships for St. John residents, many of whom worked as janitors and foundry workers in the Buick plants.

Although the area had become more racially and economically diverse, it started to be known as a poor and dangerous area when compared with the southern and eastern sections of the city. Due to its proximity to the Buick factories, which flooded the area with soot and ash, it was polluted, overrun by a significant rat population, and marked by its stock of crumbling houses. According to a 1954 survey by the Urban League, Black Americans living in neighborhoods like St. John also lacked access to reliable public services. Some 56 percent of Black Americans, according to this survey, lived with people they weren't related to, while almost 70 percent of these living quarters lacked private bathrooms (this area encompassed the St. John and Floral Park neighborhoods as well as areas along Saginaw Street).[14] The pollution was omnipresent in St. John and neighboring Oak Park and Martin-Jefferson; industrial waste dusted automobiles, windows, clothing, and pets. Due to these respiratory contaminants, area residents posted the highest rates of infant mortality, lung and throat cancer, and asthma in the city.[15]

Despite increased demand by Black Americans for homes in more diverse areas of the city, even those who managed to save for a home could not overcome the institutional barriers they faced, which included obstacles to securing financing and the local practices and social norms that restricted their housing choices. Flint's Black residents were stuck living in neglected and deteriorating segregated neighborhoods littered with uninhabitable structures and vacant units, with some neighborhoods even lacking sidewalks, as well as proper water and sewage services. Homes in these neighborhoods also had a reputation for being unsanitary and overcrowded.

However, housing conditions were no better for Black Americans living in city-owned public housing projects. These too were in deplorable condition, often infested with rats and roaches and eventually requiring extensive renovation. Residents of public housing projects also lacked access to properly paved roads and regular garbage collection. Even requests for identifying unit numbers in public housing projects were oftentimes denied. These spaces seemed to exist solely so that city leaders could claim to be doing something to provide low-cost housing to people who had previously been displaced by urban renewal projects. While the passage of legislation in 1966—including Senate Bill 1054, which required displaced families to be provided with standard housing, and House Bill 3781, which required those displaced by highway construction to be placed in suitable housing—was a step in the right direction, these public laws lacked necessary enforcement provisions (much less protections against discrimination).[16]

Consequently, living conditions and public services available to residents continued to deteriorate. Frustration among Black Americans regarding the persistence of substandard housing increased. This boiled over when Flint's Black residents joined those of over 150 other urban areas across the United States in civil unrest that challenged Jim Crow restrictions and housing conditions during the "long hot summer" of 1967.[17]

Residents of Flint's Black neighborhoods filled the streets on the corner of North Saginaw and Leith Streets and near South Saginaw and St. John Street on July 24, a day after civil unrest had begun in Detroit (which ultimately led to five days of rioting in that city that claimed over forty lives and caused millions of dollars' worth of property damage). In Flint, a group of instigators began throwing rocks and bottles at passing vehicles near an East Stewart Avenue grocery store. This escalated to torching establishments with reckless abandon.

While some rioters avoided looting businesses with windows labeled "Soul Brother," other establishments took direct fire and sustained considerable property damage in the predominantly Black areas of the city. Henry Sergis, owner of the Shorthorn Meat Market on North Saginaw, reported over $11,000 in damages. And even businesses that weren't torched were not spared entirely. Some were looted while others, like the People's Furniture and Appliance Co. on East 8th Street, were vandalized and storefront windows and doors were shattered.[18]

The Flint riot was relatively short-lived and neutralized by the strong presence of law enforcement. Although initial attempts by police to break up the crowds around Saginaw and Leith Streets was unsuccessful, law enforcement was eventually able to gain control within twenty-four hours. This was aided by Governor George Romney's declaration of an emergency in areas throughout the state where civil unrest was occurring, including Genesee County, and the deployment of the National Guard. Michigan State Police set up checkpoints along North Dort Highway and, with the aid of almost 250 law enforcement officers, stopped the cars of suspected rioters, barricaded neighborhood intersections, and patrolled the streets (armed with rifles) in actions meant to intimidate residents and discourage additional upheaval. Flint's first Black mayor, Floyd McCree, also imposed mandatory curfews and bans on the purchase of liquor and gasoline to help keep the peace.

On the heels of this social conflagration that resulted in the arrest of 102 Black citizens—most of whom had taken part in the civil disturbance by fire-bombing stores, smashing windows, and vandalizing vehicles[19]—residents and leaders of Flint were eager to believe that the Model Cities legislation could help solve the blight and discrimination that were plaguing their community. At the time, no other policy or community effort seemed capable of resolving the housing issues in the city's segregated neighborhoods. Flint had seemed immune to attempts by residents to bring attention to the overcrowded and dilapidated housing problems in Black neighborhoods, even after citywide campaigns were launched to advocate for better living conditions.

A decade earlier, attempts had been made to improve housing conditions by inspecting living quarters in the city's segregated neighborhoods. The owners of rental units were fined for violating building codes and cramming families into rundown homes (and occasionally basements with faulty wiring, leaking roofs, peeling paint, and rodent issues). However, these code enforcement efforts were short-lived and did not instigate meaningful improvements in Flint's Black housing stock (where over two-thirds of the residents owned their homes).[20]

In April, a few months before the 1967 riot, the Genesee County Board of Supervisors executed a resolution authorizing the Genesee County Planning Commission to draft a planning grant application to help develop a fully realized Model Cities proposal. The hope was to address housing conditions as well as mobilize local leaders and agents of change in the community. The application, however, was not successful, and Genesee County was dropped from the program's first round of cities.

After local officials made considerable effort to revise the application and address gaps in its plans for incorporating citizen input into the program, HUD accepted Genesee County's application in March 1968. The county was awarded $200,000 to develop its Model Cities program plan, including taking into account the needs of residents in the 3,200-acre area that included Flint and the townships of Genesee and Mt. Morris. Before the planning grant contract was received on August 8, major appointments to the Community Development Act (CDA) staff had already been made, and a headquarters "opened with little or no fanfare, equipped with two pop cases for chairs and a borrowed table and typewriter stand as the only office furniture."[21]

With the Model Cities program staff assembled, community aides were hired to knock on doors, locate local leaders, distribute information about Model Cities initiatives, and organize community meetings. When the county received the HUD grant (and additional financial support from the Charles S. Mott Foundation), the Genesee County Model Cities program staff budget reached almost $310,000. Nearly two-thirds was allocated to personnel and consultants ($126,000 and $70,000, respectively); about $15,000 for travel, supplies, and office space; and $95,000 for the "removal of impediments for citizen participation."[22]

Henry Horton, who was appointed Model Cities director, got right to work to ensure the program's success. This task included reaching out to Flint's residents and getting them excited about the opportunity to revitalize the city's blighted neighborhoods. For the program to work, Model Cities planners knew that community buy-in was required—not only to participate in the HUD program but also to implement equitable and sustainable housing improvements. Director Horton insisted that residents in Model City areas needed to feel a sense of comfort and pride in their neighborhoods. He set out to ensure this by making plans to shore up police protection in the area, which was over 60 percent Black American. Horton suggested that "store-front police stations" be staffed by officers with at least two years on the job and that the City of Flint consider expanding the number of Black police officers in the area (there were only 4 Black police officers on Flint's 400-person force).[23]

Meanwhile, local organizations wanted to participate. Local leaders, including Flint Urban League executive director John W. Mack, made efforts to touch base with Model Cities staff and ensure programming goals were appropriate for the area, since conditions in Flint were, he wrote, "worsening. . . . Flint may soon be a black ghetto ringed by white suburbs." He went on to say, "These are crisis conditions which will decrease the tax base within the city; increase the expenses of the city; produce more problems for the school board; decrease business within the city; and increase the tax burden upon the county, state, and federal governments."[24]

Model Cities director of housing David Leonard responded to these concerns —and skepticism from local organizations (and HUD, which also raised questions about the city's capacity to overcome existing social barriers)—by promising cooperation from local governments within the county, guaranteeing support from local civil rights organizations, and vouching for the continued private funding of the Mott Foundation. Attempting to calm concerns about persistent racial discrimination, especially in the areas of employment and housing, Model Cities staff shared in communications that banks and other lending institutions, as well as big businesses, were open to improving

housing conditions and were starting to "show more interest in problems related to slums and poverty." Ultimately, Leonard recognized that rejuvenating
and revitalizing Flint's blighted communities would not be easy, but the work
just might pay off in the end. "The general climate in the power structure and,
perhaps, in a large portion of the citizenry, is one of [doubt] and questioning,"
Leonard acknowledged. "The old values, standards, and patterns of action hang
on and [will] largely determine the course of action or inaction, but there is a
feeling that the time may be ripe for change."[25]

In addition to providing direct responses to questions and concerns voiced
by various stakeholders, Model Cities staff also encouraged local support by enlisting help from over twenty countywide human service organizations, including the Department of Social Services, the Urban League, the YMCA, and the
Police Community Relations Department. At least thirty-four meetings were
held to coordinate this support. Accordingly, to get a sense of how residents felt
about their neighborhoods, Model Cities staff created surveys that were passed
out by community aides in front of shops and throughout residential areas near
Saginaw Street. In 1968, between September 17 and November 27, the Model
Cities programming staff held over sixty-five meetings at their office with residents to discuss problems and develop improvements.

In addition, the Model Cities Arbitration Board was positioned to field complaints from residents and settle conflicts or grievances with decisions made by
the Genesee County Board of Supervisors. Technical Advisory Groups were
set up to suggest best practices to Model Cities staff, and the newly minted
Social Planning Department was created to offer guidance to task forces that
used demographic characteristics from model neighborhoods at the national,
state, and local levels. Elections were also set up to ensure that residents were
provided an opportunity to vote on the community representatives for the
Martin, Dewey, and Beecher district councils. Residents could cast their vote
for anyone over the age of sixteen. Those selected would be tasked with serving
on evaluation and planning committees and thereby positioned to "assure that
[the] measures of project and program successes which concern the citizens are
being adequately considered by CDA staff."[26]

On October 15, 1969, HUD announced its approval of Genesee County's
plan for upgrading blighted areas and incorporating citizen participation in
the planning. Model Cities activities began in Flint.[27] The Genesee County
Model Cities program was eventually awarded a block grant for just over $3.5
million for its first year, which could be used as the county wished in tackling
social problems not currently addressed by existing federal antipoverty programs. Eighty percent of the funds awarded to Genesee County during the
Model Cities years was spent in Flint (about $1.5 million a year). The programs

targeted a blighted area in Flint that made up only 14 percent of the city's area yet accounted in 1967 for over 30 percent of its crime, 75 percent of it committed by individuals under age twenty-five.[28]

A seventy-two-member Flint resident council was created, consisting of fifty-seven members elected from districts throughout the city and fifteen youth representatives elected at large. It was charged with developing Model Cities plans by identifying potential concerns, ranging from the lack of libraries on the north side of the city, to newly constructed houses falling apart only a few months after completion, to human services agencies failing to respond to community needs.[29] Council members were assigned to subcommittees on housing, education, employment, municipal services, health, social sciences, transportation, public safety, and recreation. These subcommittees were delegated with drafting problem statements and suggesting ways to overcome communal issues. Unelected Flint residents involved in the Model Cities effort volunteered to help clean up the community, while some youth were employed to assist in these efforts.

Once the Model Cities initiatives swung into action, residents in targeted neighborhoods were able to take advantage of a variety of programs. For instance, old cars and debris were removed from vacant lots as a result of the Bulk Pickup program. Residents were educated about environmental health hazards and ordinances through the newly established Environmental Health Outreach program. As part of the Neighborhood Improvement program, 25 youth workers spent a summer trimming dead tree limbs and discarding items that cluttered vacant lots—old televisions, stoves, garbage, dead animals, and abandoned car parts. Another 200 of Flint's young people had an opportunity to work with Genesee County Parks as part of the Summer Employment Youth program, known as SEY 200. Working in twenty crews with 10 people each, these young people were tasked with cleaning and maintaining county parks as well as the Flint River's forty-mile shoreline (from East Carpenter Road to the Holloway Reservoir).[30]

Model Cities efforts also set up Girl Scout troops during the summer in "hard-to-serve areas" and developed training programs known as CAMPS (Coordinated Area Manpower Planning Systems), which helped unemployed women secure clerical or sales training and job placements with local employers (including Smith-Bridgman; Sears, Roebuck, and Co.; and Flint Fabric). Contests were created for residents to select Model Cities insignia that would be printed on neighborhood patrol uniforms, with a $50 savings bond offered for first place and $25 for second place. Additionally, Model Cities residents were promised a role in selecting their own police force. With $194,000 from the Metropolitan Police Program, residents were informed that police officers who had been recommended by Flint police chief James Rutherford and Genesee Township police chief Lloyd Goudy would be hired. They were also assured that "final approval [would] be in the hands of the citizen's committee."[31]

Finally, Model Cities program funds were used to help poor residents, like eighty-year-old "Mrs. Carrie," secure suitable housing. After Carrie's deteriorated home on Baltimore Street was condemned and slated for demolition, she received help from the Model Cities Household Assistance Program representative and, four weeks later, was able to secure a "home with a living room" in a modern Forest Park manor apartment. For many low-income Flint residents relying on Old Age Assistance, support from this particular Model Cities Household Assistance Program was especially welcome. Without such aid, including the deposit and first month's rent, few Flint residents could escape the perils of aging homes and displacement by public action.[32] Citizen-controlled programs like the Model Cities Demolition Corporation also helped accommodate residents who lost their homes and were seeking safe and stable housing financed under the FHA.[33]

When the initiatives and groups created during this time through the Model Cities program proved unable to resolve a problem, community groups often sprang up to attempt to address the issue. For instance, when prolonged advocacy failed to secure a public commitment to build a library on the north side of Flint—one of the city's most diverse and impoverished areas—representatives from various PTA, community council, and Early Childhood Development groups formed Concerned Citizens of the North-End of Flint. This group persuaded the board of education to adopt a resolution providing north-central Flint a public library—*if* a proposed millage was passed, operational funds became available, and appropriate space was found. The millage passed in March 1972, but the Concerned Citizens of the North-End never got their library. The board of education, pointing to the seven public libraries it already operated in Flint at this time, opted out of building the proposed library—even after this strong community effort on behalf of the 48 percent of North End children who lacked access to a public library.[34]

Persistent Challenges to the Model Cities Program

Privileging Middle-Class Goals while Neglecting Working-Class Realities

Since in some areas—like Benton Harbor—Model Cities plans were at a standstill, initiatives in Flint appeared successful. Programs were created; citizens were elected to serve; salaries were paid. Rules, expectations, and goals were laid out and broadly distributed. But the bad vibes between residents and city officials remained. A significant percentage of the homes available to Black Americans within the model neighborhood stayed vacant and uninhabitable. And the housing crisis in Flint continued to be compounded by widespread unemployment and poverty.

The archives that I reviewed at the Genesee Historical Collections Center show that there were early warnings that the Model Cities program would never live up to its potential. Model Cities proposals, like Community Partnership in Action, stopped hiring youth and began leaving positions vacant— until the chair of the Joint Citizens' Council insisted that the organization use available funds to hire youth. Residents, including those involved with the program as CDA and Neighborhood Center staff, began sharing their frustrations concerning how the programs were running with the district executive of the Michigan Civil Rights Commission and acting executive director of the Community Department, Mrs. Olive Rankin Beasley, who began her career processing employment discrimination claims as a field representative for the Michigan Fair Employment Practices Commission.[35] While various issues impacted the quality of Model Cities plans in Flint, including the high turnover of CDA staff, the most significant issue was the fact that the programs actually better served the needs of the middle class and not the poor, the original target population. Instead of using funds set aside to improve housing conditions, public accommodations, police community relations, and health disparities, Beasley noted this in her August 1970 quarterly report just one year after Genesee County's Model Cities proposals were approved: "Funds were going out [but] the economic development within the target area remains at status quo."[36]

Making matters worse, despite the importance of citizen participation in this HUD program, it was clear that efforts to involve local residents in the planning and execution of Model Cities goals were quickly losing steam. District councils were not allowing the citizens elected to represent the interests of their constituents participate in the decision-making process. The councils were being co-opted by local professionals serving in advisory roles who did not represent the interests of people in targeted areas. According to Beasley, district councils had developed into "closed system[s]" that had been overtaken by the creation of an "establishment within . . . Citizens Council[s] which approximates the power structure in the general community." This takeover of citizens' councils had a ripple effect within Flint's Model Cities program. The number of eager volunteers began to dwindle, and some staff even stopped showing up to work because of their "disillusionment" with how these programs were being run.[37]

If the needs of residents contending with a shortage of suitable housing were indeed being considered, program funds should have been used to build homes instead of to demolish structures. However, the opposite was the case. By October 1974, demolitions substantially outnumbered units constructed: 492 homes had been torn down while only 12 housing structures had been built.[38] Residents, like those dislocated from the Doyle and St. John neighborhoods after acquisition and demolition, found themselves relocated to neglected and predominantly Black areas in northwest Flint. Over 87 percent of citizens in these neighborhoods were racial minorities who had the lowest incomes in the city.[39]

The priorities of the Model Cities program in Flint were also shaped by how the city spent HUD funding. Between 1968 and 1972, the city received about $8.7 million in HUD funding for Model Cities initiatives, yet only 18 percent was spent on public service programs. Among activities funded, capacity-building programs received more funding than those designed to rehab neighborhood spaces. Education and recreation/culture programs received 23 and 21 percent of funding, respectively, while crime prevention received 9 percent and environmental protection/development programs 2 percent. In Flint's 1973 budget, no funding was allocated to public service programs. Instead, the city indicated that $703,300 of its "contingency" funds budget might be spent on public services.[40]

It turns out that the allocation of these contingency funds was never made. Public service programs in Flint continued to be unfunded until 1976, when public services made up just over 3 percent of the city's $8 million Community Development Block Grant funding. The public service programs that received funding supported small business services ($59,901) and cultural enrichment ($227,638). The vast majority (nearly 80 percent) was allocated to programs and activities connected to the city's urban renewal initiatives, including property acquisition, demolition, and relocation payments.[41]

Entrenched Bias in the City of Flint

RACE AND FLINT SCHOOLS

Another factor that undermined Model Cities program goals was Flint's extensive history of widespread racial prejudice and discrimination. This shaped all areas of communal life. Due to the significance of race, white and Black families continued to live separate realities before and after the Model Cities era. They lived in segregated neighborhoods, they worshipped in segregated churches, and their children attended segregated schools.

The Flint Community Schools district did commit to a program of cultural enrichment that featured performances by dance troupes, drummers, narrators, and actors, and it made an effort to incorporate Black studies and other ethnic group materials into a specialized curriculum.[42] However, the school district also failed to implement plans to address its role in maintaining de facto segregation in Flint's schools, which included encouraging unfair suspensions and physical attacks and institutionalized hiring restrictions that severely reduced the number of Black teachers and administrators. Despite attempts of community groups—including the local NAACP support of busing initiatives, Project Onward and Upward's efforts to increase children's exposure to models of Black success, and the Flint Education Coalition's attempt to get citizen representation on the board of education[43]—racism and discriminatory administrative practices created inadequate instructional programs and an environment that negatively impacted the quality of education available to Black students in the city's schools.

In 1975, segregationist practices within Flint schools were uncovered by federal investigators. In August, the Office of Civil Rights (OCR) determined that racial bias shaped employment and pupil placement decisions in the Flint Community Schools system. Evidence secured between 1968 to 1975 supported the OCR's conclusion that the district was noncompliant with Title VI of the Civil Rights Act of 1964, which prohibits discrimination based on race, color, or national origin in any program receiving federal funding. In the determination report, the OCR noted that it discovered "racially motivated policies and practices with respect to the assignment of students and the employment of faculty and other staff." Not only were there enormous racial imbalances in student populations, but there were enormous racial imbalances in the teaching assignments within the District. The district had avoided hiring qualified Black teachers before 1943 and had continued to assign Black teachers and staff to predominantly Black schools. For instance, in 1965–66, 53 percent of Black teachers were assigned to Northern and Northwestern High Schools (33 and 45 percent Black American, respectively). Meanwhile, 42 percent of Black teachers were assigned to schools like Central and Southwestern, whose students were 20 and 18 percent Black American, respectively. By the 1974–75 school year, 70 percent of the 102 Black teachers employed by Flint Community Schools would be assigned to predominantly Black American middle schools.[44]

Student enrollment remained subject to racial bias.[45] For instance, 82 percent of Black elementary students were directed to enroll at eighteen schools that had Black student enrollments at 56 percent or more, while the largest predominantly Black high schools, Northern and Northwestern, housed 72 percent of Black high school students. The OCR also determined that, in the decade between 1960 and 1970, Flint Community Schools had systematically assigned three times the number of teachers who were not fully certified to primarily Black schools rather than mostly white schools. This meant teachers assigned to predominantly white schools had more education and more teaching experience, while the OCR found that predominantly Black schools got the least credentialed teachers, one factor that resulted in lower student achievement scores.[46]

The OCR report led to desegregation efforts, and *those* led to a host of negative repercussions. Just after the federal government ordered Flint schools to desegregate in 1975, the Mott Foundation withdrew its financial support for community education initiatives, leaving the Flint Board of Education to deal with the challenge of integrating racially segregated neighborhood schools.[47] Racial tensions were so raw at this point that the Flint Board of Education and the Genesee Intermediate School District ignored the federal order to desegregate. Later, community groups, including the Flint Black Teachers Caucus, and concerned parents (including Barbara Cook, Sondra Bell, and Mary L. Miller) would attempt to compel the district to desegregate by pursuing a lengthy and costly class action lawsuit filed on March 23, 1976. In their complaint, the plaintiffs maintained that the board of education never created a desegregation plan and continued to maintain a separate and unequal public education system based on race and color—a system that created learning disparities among the area youth. The local NAACP and community groups, including the Concerned People for Quality Education, supported the suit. Yet it dragged on. By 1978, Flint's elementary schools remained racially segregated, with 82 percent of elementary students attending segregated schools and thirty of the thirty-seven schools "racially identifiable" despite litigation threats from the Department of Health, Education, and Welfare.[48]

Ultimately, the lawsuit was dismissed by presiding judge Stewart A. Newblatt. The 1980 decision claimed that plaintiffs had not proven their "non-specific allegations" and theories of racial discrimination within Flint Community Schools and neighboring suburban districts.[49] It would take a consent decree, signed in the spring of 1980 by board of education officials and the Department of Health, Education, and Welfare, to force the school board to expand its elementary school magnet program (in an effort to improve interracial enrollments). White families now began to avoid the city altogether. The proportion of white students in its public schools decreased from 53 to 30 percent between 1980 and 1992.[50]

RACE AND PUBLIC SAFETY

During the Model Cities era, neighborhood or "store-front" police stations sprang up around Flint. Images of community police conversing with residents and of officers on patrol were frequently captured in the *Model Cities News*, the program's monthly periodical that claimed to "Tell It Like It Is." Despite these activities, goals for improving public safety and police–community relations in Black areas were not realized. A strained relationship between Flint citizens and law enforcement continued due to a long history of harassment, violence, and intimidation.

A month after Flint's 1967 riot, a few residents representing the Committee of Concerned Citizens tried to intervene on this issue. They enlisted the Michigan Civil Rights Commission in an effort to address allegations of misconduct toward Black Americans by police officers. In response, the commission offered three potential solutions. First, it noted that it could intervene and investigate but warned that while it "has the power to issue cease and desist orders, [it] does not have the power to issue orders of a punitive nature." Second, the commission suggested public hearings but cautioned that this option "might provide more heat than light on the subject" and might be overtaken by attention-seeking individuals "in the glare of publicity . . . to make pronouncements and statements rather than to give information." Finally, the commission strongly recommended its third alternative, which was a study of police–community relations that would document facts in an "independent, scientific manner." However, such a study never materialized, and the concerns about police brutality from Flint's Black residents were never addressed.[51]

The Genesee Historical Collections Center archives are full of complaints about police intimidation and allegations of police abuse during the era. Flint residents typically called upon Edgar Holt, president of the local branch of the NAACP, for assistance after having been roughed up by police. For instance, a month before HUD awarded Genesee County its planning grant, Troy Mitchell reached out to Holt after his brother, Walter Mitchell, was physically assaulted by the Flint police. NAACP staff phoned Mitchell at the hospital to learn more about his complaint. In this phone conversation, he shared that he had been arrested for intoxication near Berston Field House. Despite his willingness to enter the patrol car when requested, Mitchell said he had been beaten after being handcuffed by the arresting officers. Holt reported that "they called him nigger several times and beat and kicked him into a cell and beat him some more, after that, they took him to the hospital for treatment. The hospital medical records show that he entered the hospital for treatment of a lacerated scalp."[52]

The NAACP staff learned that there were two witnesses to this abuse. Mildred Jenkins saw what was happening when traveling north on North Saginaw (she was grabbing something to eat with her companion, Theophus Griffin, after catching a movie). When she noticed that someone was being beaten in a police car, she said that she pulled over into the Summerfield Chevrolet parking lot to watch the situation play out, mentioning that she saw "policemen kicking the man in the lower part of his body. Another police car arrived, and apparently, without asking any questions, another patrolman began to kick the man and slammed the car door on his limbs." Griffin's account dovetailed with Jenkins's.[53]

Another example of police intimidation manifested just before 11:00 P.M. on Sunday, June 29, 1969, when six police cars converged on a home at the corner of North and Wood. Officers jumped from their cars, guns in hand, in response to a report of a gunshot on McFarland Avenue—about three blocks south of this neighborhood. An eyewitness, Ray Pallas (who had credentials to enter a blockaded area), reported to Thomas E. Sagendorf, director of the Interfaith Action Council of Greater Flint, that the police officers entered the home, occupied by a Black man. They pushed the homeowner aside, briefly searched his house—without a subpoena or getting his express permission—and refused to say what they were looking for. (The man would later learn their purpose from a neighbor.) The officers then went to another home but were again unable to find the weapon that had been fired or the person who allegedly had fired it. Then, Pallas reported, the police began to cruise the area, speeding down block after block with their lights flashing. Nearly thirty Black residents had gathered in the streets, and none claimed to have heard any gunshot. They were deeply disturbed by what felt like a police invasion—and they were not calmed by the arrival, the next morning, of plainclothes officers canvassing their homes and warning that they should not talk about the event.[54]

A few months after this incident, William Brown, an area bus driver, contacted the NAACP after a December 24, 1969, encounter with the Flint police. While crossing the street, Brown had spotted children whom he knew and stopped to wave at them. After observing Brown pause, Flint police officers jumped out of their car and grabbed him by the wrist (which was being treated for an injury at the VA clinic). Brown's complaints of pain did not win any sympathy from the officers, and in response to his protests, he recalled that one of the police officers said, "Nigger, you don't have any say so ... I couldn't care less about your condition," and then struck him over the head. After they put him in the patrol car, Brown said that his wife asked the officers, "Where are you taking him?" One of them replied, "Downtown, and if you want to go, you can jump your ass in."[55]

Brown was booked and alleged a few more police officers attacked him, claiming they jumped him, struck him on his back and head, and eventually broke his arm. When officers noticed that his arm was broken, Brown recalls someone saying, "We'll have to take him to Hurley now." Due to his injuries, a four-day hospital stay was required. Upon discharge, Flint police escorted Brown back to jail in his hospital clothes and, upon arrival, informed him that he could leave by paying twenty-five dollars. Brown called his wife, who bailed him out and took him home in his hospital gown—after finding out that the police had misplaced his clothes.[56]

By the early 1970s, Flint residents were still trying to do what they could to improve police–community relations in Black areas. This effort was an uphill battle, since ranking officials within the Flint Police Department—as well as then mayor James W. Rutherford, formerly the Flint police chief—categorically denied residents' concerns with what they called "over-kill syndrome" or police brutality.[57] City residents created ad hoc committees for different issues to get answers and to advocate for solutions. They also participated in community groups, including the Interfaith Action Council (a collective of white and Black residents) and the Michigan Catholic Conference, which called for a legislative hearing on ongoing police-related controversies in Flint. By the time interest in this proposal for legislative attention fizzled out, Citizens Resources—a community organization formerly known as Citizens for Better Police Community Relations—started making its own attempts to address police–community conflicts in Flint.

After holding several meetings throughout 1971, Citizens Resources began making consistent requests for information pertaining to complaints about Flint's police. Group members wanted to know the number and the status of pending complaints, which usually were for false arrest (such as being picked up on suspicion of public intoxication) and excessive force allegations. They also wanted to gain access to rules governing the Flint Police Department's complaint procedures.[58] Instead of resolving such cases and clarifying the process, efforts were made to discourage and intimidate people who might file a police complaint, including by the Fraternal Order of Police—which represented law enforcement officers in Genesee County—which announced in late 1971 its intent to sue any person for libel or slander who filed a complaint based on allegations or evidence not admissible to courts.[59]

Citizens Resources' request for information on police complaints was a continuation of previous inquiries in 1969 into such matters by Edgar Holt on behalf of the NAACP and by Burton I. Gordin, the executive director of the Michigan Civil Rights Commission. Gordin had sent Walter Greene to meet with Olive Beasley and Holt regarding the unresolved complaints. Greene assured them that the resolution of the cases would be expedited. Nevertheless, complaints continued to pile up against officers accused of violating the rights of Black Americans with excessive force and false arrest.[60] Despite all efforts to address the police department's reputation for intimidating and harassing members of Flint's Black neighborhoods, the police department seemed above reproach. It sat on reports and generally refused to comply with documentation requests or to provide updates regarding the status of pending cases.[61] By the end of the

Model Cities era, the Flint Police Department was known for not cooperating with internal abuse investigations or with those by the Michigan Civil Rights Commission and, instead, forced complaints to be referred to the attorney general's office for review.[62] Many cases had been pending since the notorious racist George Wallace's 1968 presidential campaign appearance in Flint.[63]

"New Federalism" Takes Over with a Period of Benign Neglect

The civil rights movement increased the federal government's involvement in social spending.[64] Then, in the early 1970s, Richard Nixon's "New Federalism" initiated a significant shift in intergovernmental fiscal relationships. Nixon opposed spending on categorical programs, preferring block grants and revenue sharing (in the name of returning power to local communities).[65] Once Flint started benefiting from Community Development Block Grant (CDBG) funds, the city used them to assist census tracts with high concentrations of low- to moderate-income residents in developing city-building strategies for locating low-wage jobs. But for residents in the deteriorating, predominantly Black areas of Flint, neighborhood conditions continued to decline as program funding lagged. Minimal funds were made available to these residents.[66]

Flint's uneven and discriminatory neighborhood rehabilitation plans are perhaps best illustrated by the St. John Urban Renewal Project that created Industrial Park, the first area project funded by the CDBG program. The city leveraged an Urban Development Action Grant from HUD to acquire a large piece of land near Buick City, clearing the area and making improvements meant to usher in new businesses. Other CDBG funds were used to acquire and eliminate deteriorated neighborhoods. However, conflict emerged quickly as community groups and civil rights organizations became concerned that the city continued to relocate Black Americans to northwest neighborhoods. Although relocated residents generally saw improvements in their housing quality, Black families were moved solely to primarily Black or racially transitional neighborhoods already in decline. Research from the Evidence for Community Health Organization (ECHO) program confirms that areas in Flint that predominantly housed Black Americans during this period tended to be especially deteriorated and overcrowded. Some 16 percent of homes in the city were technically deemed deteriorated and 4 percent of homes were vacant. But in areas that were 75 to 100 percent Black, 41 percent of the properties were dilapidated or deteriorated, 57 percent were vacant, and 51 percent of vacant homes were inhabitable.[67]

While ignoring the needs of poor people, the City of Flint used CDBG funding—almost $30 million by 1975—to support initiatives that would enhance the mobility of middle-class residents, including economic development loans to minority- and women-owned businesses. After the Flint Neighbor-hood Improvement and Preservation Project incorporated—originally cre-ated in 1976 by Mayor James Rutherford with a mission to improve conditions in poor neighborhoods—it eventually changed its focus to rehabilitation projects in moderate-income communities. The project continued to make some inroads in poor communities by supporting neighborhood improve-ment initiatives and founding more than 200 block clubs,[68] but infrastructure improvements—including the installation of new curbs and gutters and street resurfacing—were primarily reserved for middle-class communities in Flint.

Remaining CDBG funds were earmarked for urban planning initiatives meant to revamp the city's reputation. For instance, planners committed CDBG funding to projects designed to create a tourist industry in Flint, includ-ing the development of Autoworld, Waterstreet Pavilion, Windmill Place, Car-riage Town, and the downtown Riverfront Center Hyatt Regency—ventures that would eventually fail to inspire any "renaissance" of Flint's city center or create new jobs for low- and moderate-income residents.

Although Flint committed considerable funding to relocate residents from acquired properties, residents and officials within the Michigan Civil Rights Commission complained that relocation services were inadequate. In a June 7, 1973, letter to Richard Wilberg, executive director of Flint's Department of Community Development and Urban Renewal Administration, Olive Beasley noted that Flint needed to restructure and redesign relocation services so that they were "compatible with the intent of the U.S. Housing Act Urban Renewal provisions . . . which is to improve the quality of life." According to Beasley, "Flint has an obligation to persons it is displacing to accommodate the needs of the major industry and to compensate for the neglect by the City that has created the blight and deterioration of slum areas." Beasley urged Wilberg's de-partment to develop a comprehensive relocation system that reduced the stress of those displaced and would help "restore public confidence in the integrity of city government, particularly in the minority community."[69] Unfortunately, these accommodations never truly emerged.

Consequences of Racial Prejudice and Residential Segregation in Flint

Although optimism about Model Cities initiatives and antidiscrimination legislation, as well as the funding from the Charles Stewart Mott Foundation, flowed through the city, Flint could not escape its attachment to residential segregation. While it is true that Flint was the first city of its size in the United States to pass fair housing legislation with Ordinance #2008—on February 20, 1968, by the markedly slim but prevailing margin of 1 vote[70]—it turns out that this did not change the opinions of those who vehemently opposed desegregated housing (just as their minds hadn't been changed about desegregated schools). Neither did the federal open housing bill that was passed by Congress in April 1968. So, rather than benefit from urban renewal programs designed to improve older neighborhoods, Flint's Black Americans became overwhelmingly concentrated in its declining northwest neighborhoods. Housing covenants, city ordinances, and high home prices rendered the southern and eastern parts of the city off-limits to most Black families.

No doubt overt actions undertaken by white supremacist organizations and blockbusting real estate agents helped perpetuate residential segregation.[71] Realtors, in fact, established a practice of helping neighborhoods remain racially exclusive. Evergreen Valley, a subdivision located on the city's border with Burton Township, had been developed as an all-white community in the 1950s. When a property in the area was sold to a Black American in July 1963, it became the subject of blockbusting tactics. Local realtors started visiting homes on integrated blocks and encouraging white homeowners to put their properties on the market. It helped turn areas that were desegregating into hotbeds of white flight.[72] And the trend would continue, at least in this neighborhood, until the citizens of Evergreen Valley formed the Evergreen Valley Association, which worked to stop panic-induced sales tactics designed to scare white homeowners away from living near Black Americans.[73]

This legacy of racial segregation in Flint made things more difficult for the Model Cities initiatives that sought to reinvigorate the city. Most of Flint's low-income Black residents, for instance, were concentrated in blighted communities or city-administered housing projects, where residents paid between fifty and seventy dollars per month—a significant share of their monthly income—to live in rat- and roach-infested, poorly maintained dwellings. The sidewalks crumbled; garbage piled up. Even house numbers were missing. Signs of neglect and improper maintenance were evident throughout the community. However, in Flint proper, less than 1 percent of homes were uninhabitable, and around 81 percent of the dwellings were well maintained. It was clear that housing conditions were deplorable for Black Americans; they occupied 24 percent of the housing units in Flint, but predominantly Black areas had more vacant properties and deteriorated units than parts of the city deemed transitional (30 to 74 percent Black[74]) and white, based on the racial composition of these spaces.[75]

The Ultimate Opportunity Squandered

Genesee County hosted the most developed and innovative Model Cities plans in Michigan. In Flint, Model Cities funded and coordinated police patrols, influential citizens' organizations, and social service activities. However, the program did not live up to its potential, primarily due to the continuing significance of race and the program's transformation into a middle-class intervention effort that did not do enough to improve the structure of opportunity. Residents were organized and promises were made. But the politics of race and targeted investment prevailed. As a result, this opportunity to create equitable and sustainable improvements in Flint was squandered. As Olive Beasley put it, although Genesee County had "82 projects approved and many [were] in some phase of execution . . . most of the projects represent[ed] merely an extension of existing community services and follow[ed] the traditional pattern of taking money out of the black community."[76] Even as block grant funding began pouring into the city, Flint residents—especially in Black areas—continued to complain about the quality of their public services and to endure the deteriorating conditions that remained.

The Model Cities program presented an opportunity to Flint. Residents rightly hoped it would mean that their complaints about essential services could be addressed. Yet long-standing segregationist tactics and class politics undermined these efforts, just as they did in other cities that sought to intervene in the legacy of residential segregation and its ecological consequences.[77]

At the same time, organizations of "concerned" citizens created during this era survived to fight for social justice and equality in Flint. The streetlights neatly placed on each lawn in some struggling communities is a show of solidarity, a physical reminder of hardworking community members who never ceased attempting to improve and guard their space. Everything *else* about Flint's poor Black areas, on the other hand, is a reminder that even the savviest, most dedicated community group cannot, on its own, overcome the long succession of bureaucratic hurdles and interpersonal conflicts that obstruct the path toward equitable and sustainable growth.

2

The Ecology of Dreams Deferred

Neoliberalism — Wrecker of Environments —
Comes to Flint

"I believe that today, July 4, 1984, is the first day of the rebirth of the great city of Flint," declared then governor James Blanchard, as 12,000 balloons were released into the sky at a pep rally in front of thousands of onlookers.

Citizens and marching bands were on hand not just for the national holiday; they were also commemorating the opening of AutoWorld, the city's new $80 million indoor theme park. The assumed success of the venture led the incoming governor to be bullish on current conditions in Flint changing for the better.

"Mark my words," he said, "we will look back on this day with pride because we are seeing a resurgence." He went on to proclaim that Flint would be known as "the comeback city of America."[1]

An estimated 6,500 people had flocked to Flint from around Michigan—as well as surrounding states—to take part in AutoWorld's opening, waiting in line anywhere between forty minutes to an hour for entry. Lured by a mass media advertising campaign, feature articles in the *Detroit News,* and speculative word of mouth, people were excited to visit what commercials described as "a totally new world . . . with something for everyone." Visitors would have a chance to catch the IMAX film *Speed* on the six-foot screen, while others could flock to attractions like the fifty-five-foot Ferris wheel or tour the automobile history exhibits.

Promised to be a place where "they could leave the real world behind," the lure was short-lived. The park briefly closed six months after opening day once an investment firm—based in Rockville, Maryland—pulled its funding due to poor attendance. Instead of igniting a new day in Flint, the closure of this high-profile attempt at a new beginning was plagued by operating problems far too similar to those in the real world. AutoWorld's failure would help trigger another downward spiral in the city—a sign of things to come.[2]

I always thought that AutoWorld failed because the developers couldn't decide whether it would be a theme park or a museum. As a compromise, it became a horrible manifestation of both: a museum for an automobile industry that was in decline dressed up with low-thrill rides that failed to amuse. With attractions like a three-story-tall V6 engine, a robot dressed as an autoworker (speaking of things to come) who worked on an assembly line while singing "Me and My Buddy," and a park mascot named "Backfire the Clown," it's obvious from the rearview mirror that AutoWorld was destined to fail.

To make matters worse, the rides frequently didn't work. A few days after opening day, rides like the "Humorous History of Automobility" starring Fred the Carriageless Horse, "The Great Race," and the vintage 1916 carousel were no longer functional. AutoWorld again closed temporarily just two years after opening and permanently in January 1991.

This saga compromised the city's downtown revitalization efforts, which also included the Hyatt Hotel that had opened in 1979 and Windmill Place, the $3.7 million small-scale marketplace and eatery that had opened in 1981. In the end, AutoWorld, a venture heavily funded by the Charles Stewart Mott Foundation, was designed to lure additional commercial activity to downtown Flint and benefit the retail business establishment. City planners and private investors hoped to rebrand the city as a tourist destination. Although Flint's reputation took another hit when AutoWorld closed, the park's failure had far less effect on everyday Flint residents. They were busy facing more important issues, including a local labor market dominated by factory jobs that were dwindling, the threat of eviction due to their limited circumstances, and the city's need to drastically reduce public services because of budget shortfalls.[3]

Flint's residents were also dealing with the impact of increasing drug use. Crack cocaine, known for its intense and instant high, swept through poor communities of color across the nation, peaking between 1982 and 1985. Then a massive cocaine underground economy blossomed in Flint during the mid- to late 1980s, leading to an illicit employment boom and the criminal charges that followed.[4] The Flint Police Department's Special Operations Unit had eighteen full-time narcotics officers in 1989 and made 2,393 arrests from raiding 281 drug houses (overwhelmingly crack-related). Law enforcement officials stated that about 70 percent of serious crime in the city stemmed from crack cocaine.[5] As the realities of the drug trade swept through Flint—the dealing, the prostitution, the homelessness, the petty crime—lives were upended, families were devastated, and many of Flint's neighborhoods came crashing down.

Public spaces in poor areas became littered with dirty needles, syringes, and crack vials. The risk of being a victim of property offenses skyrocketed. My family's home, like many others in Flint, was frequently burglarized. Residents had to display extreme willpower just to guard their domain against neighbor-hood criminals—who seemingly grew bolder by the day. Non-criminals had to take the steps necessary to protect their loved ones and property. Amid this chaos, combat, and economic loss, deindustrializing communities like Flint and nearby Detroit became marked by the moniker "Murder Capitals."[6]

There were residents and community organizations who worked hard to redirect these trends and protect Flint's kids from unwarranted harm and neglect—in effect, fighting for their future. Many poor communities were devastated by budget reductions in the post–Model Cities era. Michigan's Department of Public Health saw a $24.2 million reduction in revenue between July 1982 and January 1983. Funding for maternal and child health programs—including Medicaid screening; Women, Infants, and Children (WIC) services; family planning (Title X); and programs designed to support perinatal intensive care and services for disabled children—decreased by nearly $7 million.[7] As unemployment sent huge numbers of Flint's residents scrambling for public and charitable assistance and poverty and neglect were leaving them to navigate increasingly dangerous areas, these social programs with underfunded budgets had little hope of meeting the significant demand they garnered.

Mantras like "You can do it, you can do it, if you put your mind to it" were promoted within schools. Forward-thinking adults and community groups created enrichment programs and skill-building opportunities designed to help Flint's kids realize that their futures could be bigger and brighter than the stressful reality in which they lived. But what played out at home was the anxiety that came with being a member of a household living paycheck to paycheck. While educators worked to motivate youth to see beyond their surroundings, community organizations throughout the city got busy attempting to change the conditions that undermined their quality of life.

One particular community effort was aimed at improving infrastructure deficiencies, particularly the environmental hazards that were the legacy of industrial pollution and the failure to deal with it. In this chapter, I document the rewards of citizen-led efforts to get Flint to clean up the toxic materials that were physical threats to the city's residents, especially vulnerable children in poor areas. I also document how local and state officials created insurmountable challenges for community activists and undermined efforts to address the negative health effects stemming from their benign neglect of poor spaces.

The Birth of Environmentalism in Flint

After the passage of the Clean Air Act of 1970, many states and cities throughout the United States began focusing attention on the evils of pollution. The

Michigan legislature passed laws in 1972, signed by Governor William G. Milliken, that were designed to control air and water pollution and required polluting industries to pay fees to cover the cost of surveillance. The following year, the state legislature—led by Republican senator Robert W. Davis—created the Department of Environmental Quality, which was charged with coordinating state efforts to reduce and control pollution.[8]

These efforts did not put an end to industrial pollution in Flint—or in the rest of the state. Michigan was slowly becoming one of the nation's leading solid waste dump sites. By the end of the 1980s, it had the nation's third-highest concentration of Superfund sites and was annually taking in nearly 4 million tons of hazardous waste from other states, resulting in it having the fourth-highest accumulation of toxic waste in America.[9]

Local companies, including General Motors, continued to dump waste into rivers and pump toxins into the air. Throughout the 1980s and the early 1990s, at the urging of groups like the Natural Resources Defense Council and the Sierra Club, the state would attempt to hold General Motors accountable. But GM factories like the Fisher Guide Plant (on Coldwater Road in Genesee Township) operated in violation of waste discharge laws. This facility, in fact, accrued an extensive record of violations related to its federal permit to discharge treated wastewater, fly ash, and stormwater into the Hughes Drain (which empties into the Flint River). By 1981, the Department of Natural Resources (DNR) had begun pressuring the automaker to comply with regulations and cited the Fisher Body Plant for 235 violations in just three years (between January 1983 and April 1986). The DNR and federal Environmental Protection Agency's monthly and quarterly assessments during this period illustrate that the Fisher Body plant exceeded standard copper levels in the first five months of 1983, the first six months of 1984, and periodically throughout 1985 and 1986. Elevated amounts of zinc, chromium, cyanide, nickel, oil grease, and other substances were also discovered. After the State of Michigan filed suit against GM and eighteen other companies over such violations, the firms settled by paying fines ranging from $16,000 to $400,000 for polluting the state's—and therefore citizens'—water.[10]

By the 1980s, all of General Motor's Flint plants were designated as potentially or actively being in EPA regulation noncompliance. Faced with skyrocketing cleanup expenses and union-protected fair labor costs in the area and unwilling to invest in modernizing its Flint plants, General Motors began gradually closing local facilities. Once the largest employer of Flint workers, GM turned its back on the city—and the mess it had made (this is the subject of Michael Moore's documentary *Roger and Me*, with the Roger in question being GM's CEO of the era, Roger Smith).[11]

Addressing the Environmental Mess Left
by a Multinational Corporation

While GM was restructuring operations—and avoiding its obligation to clean up its legacy of industrial pollution—a smaller neighborhood entity was attempting to address the blight and environmental degradation of the city. The Flint Neighborhood Improvement and Preservation Project (FNIPP) created the Better Environment for Neighborhoods (BEN) program in 1980. It was designed to help improve housing conditions and address environmental problems by offering a range of services, from financial assistance to the direct removal of blight, in order to help homeowners and tenants. BEN provided programs and services to targeted areas in Flint, including Lewis Longway (census tract 14), Hurley East (census tracts 27 and 29), Grand Traverse South (census tract 29), and Martin-Wilkins (census tract 1).[12] These census tracts were not a focus during the Flint Model Cities era, which instead concentrated on census tracts 5, 9, and 4.[13]

The main difference between the target areas during the BEN and Model Cities eras was the demographic makeup of these communities. Both efforts were directed at areas with a high percentage of homes built before 1940 and a higher-than-average percentage of poor residents. However, while the Model Cities program targeted areas that were predominantly Black with low percentages of owner-occupied homes, BEN focused on areas that had both the highest and lowest percentages of owner-occupied structures in the city (for example, census tract 1 at 79.7 percent and census tract 29 at 31.9 percent). They were also primarily white, with less than 0.5 percent of Flint's Black residents (based on the 1950 US Census estimates that were included in the city's Comprehensive Master Plan in 1959).[14]

BEN's capacity to bring about solutions to environmental problems in these targeted areas depended on citizen participation and the city's own compliance with its codes and ordinances. For this system to work, residents needed to report blight—including houses in disrepair, overgrown yards littered with trash, and junk cars permanently parked in front of homes. It was BEN's job to help achieve compliance. Suffice it to say, the program ultimately did not work, because the few complaints residents made were rarely addressed due to organizational issues—including the City of Flint's minimal interest in the program and administrative issues within BEN—which led, as noted in the program evaluation report, to "little to no action . . . to solve [existing] problems" ever being taken.[15]

The potential success of the BEN program was undermined substantially by strained relations between the Mayor's Task Force, which monitored the program, and BEN staff. Although the Mayor's Task Force was required to provide a monthly report of BEN's activities, neither the mayor's office nor BEN staff communicated or met frequently, individually or with one another. For instance, by September 1980, the task force had met only once. And only two members of the nine-person task force were present for its meeting of June 22, 1981. Without the mayor's office on board, BEN lacked the official support within FNIPP, the Department of Community Development, and the Building and Safety Division. This situation violated the interagency agreement to comply with BEN's operational policies.[16] Although the city maintained an active referral system to facilitate its demolition plans and the removal of junk cars, no other referral systems were established outside of a partial one with FNIPP, the program's principal partner.

Making matters worse, it turns out that the BEN program was set up to fail from the start. Even after six months, it lacked a line-item budget, a work plan to execute goals, and a referral system to coordinate its efforts with existing community organizations. Managerial problems were at the heart of the program's issues, resulting in work activities not being executed. There were also no policies in place to control expenses for housing inspections—a central component of the program. These had been expected to cost $70 each but ended up costing FNIPP over $600 per inspection,[17] and the resulting inspection reports became notorious for being not only expensive but also incomplete (see figure 2.1).

FNIPP's working understanding of the environmental problems in these areas was shaped by the BEN program's evaluation, which was designed to help guide intervention efforts for the subsequent three years. This evaluation, which ran between January and December 1981, was created by a program administrator under the leadership of Department of Community Development director Jack A. Litzenberg. The administrator, Gregory McKenzie, "had a free hand in developing the evaluation format and the focus" in order to "avoid any biases" that could jeopardize the evidence being gathered or compromise technical guidance being provided to decision makers.[18]

In the BEN survey, environmental problems were identified and survey respondents were asked to speak to whether or not these problems existed in their communities. The environmental concerns that residents were asked to comment on primarily involved aesthetic issues like the presence of weeds, trash, boarded-up homes, and junk cars. In the evaluation report, McKenzie noted the limitation of the survey's questions and noticed that respondents actually had "a pre-occupation with other more serious (as perceived by the residents) problems. This speculation could be one of the reasons why a relatively low number of resident generated complaints were received by BEN inspectors."[19]

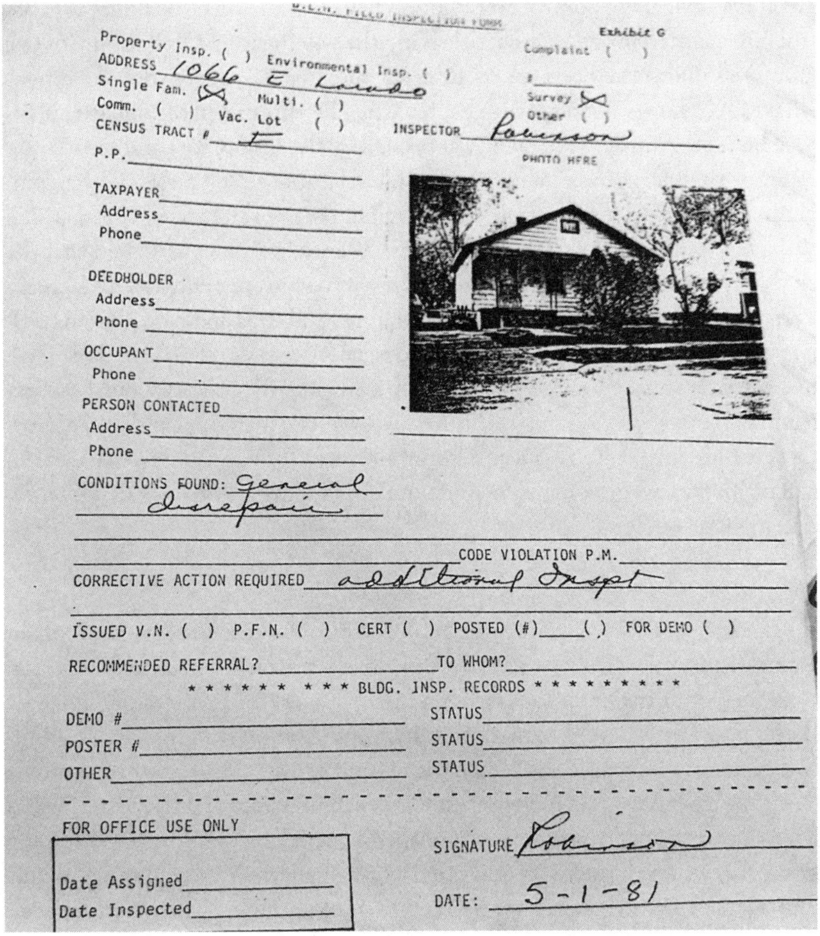

B.E.N. FIELD INSPECTION FORM

Exhibit G

Property Insp. ( ) Environmental Insp. ( )
Complaint ( )
ADDRESS 1066 E Lorado
#
Single Fam. (X) Multi. ( )
Survey X
Comm. ( ) Vac. Lot ( )
Other ( )
CENSUS TRACT # I
INSPECTOR Robinson
P.P.

PHOTO HERE

TAXPAYER
Address
Phone

DEEDHOLDER
Address
Phone

OCCUPANT
Phone

PERSON CONTACTED
Address
Phone

CONDITIONS FOUND: General disrepair

CODE VIOLATION P.M.

CORRECTIVE ACTION REQUIRED additional Inspt

ISSUED V.N. ( ) P.F.N. ( ) CERT ( ) POSTED (#) ( ) FOR DEMO ( )

RECOMMENDED REFERRAL? TO WHOM?

****** *** BLDG. INSP. RECORDS *********

DEMO # STATUS
POSTER # STATUS
OTHER STATUS

FOR OFFICE USE ONLY
Date Assigned
Date Inspected

SIGNATURE Robinson
DATE: 5-1-81

Figure 2.1. BEN program inspection report. ("BEN Evaluation Study" [memorandum by Gregory McKenzie, P. D. Administrator, addressed to the City of Flint], August 20, 1981, Cox Papers, box 3, Genesee Historical Collections Center, University of Michigan–Flint)

Problems in the Air

McKenzie did not need to speculate about the survey's capacity to capture residents' concerns accurately with regard to environmental hazards. Many Flint residents were actively committed to addressing the area's environmental challenges. They were defining what they meant by "environmental problems" and focusing their advocacy on the issues that mattered to them. For instance, Flint's citizens had been complaining about air pollution since the late 1970s.

Flint's struggle to control air pollution began when the city's pollution control office closed for good in 1976, leaving the Michigan Air Pollution Control Commission on the hook for monitoring the area. I suspect this outcome is what Mayor James Rutherford expected when he eliminated the office from his 1976 budget, a move subsequently approved by the Flint City Council.

Since the local government had chosen to not even log air pollution complaints, citizens with concerns had to reach out to the state commission, which had a habit of ignoring them. By 1978, air quality in Flint was so bad that the city was in danger of losing federal funding if steps were not taken to decrease harmful oxidants in the air that irritated lungs and eyes and contributed to allergy and respiratory problems. Then-current federal standards required that, by 1982, oxidants be limited to .08 parts per million for a twenty-four-hour period, while in 1978 oxidants in Flint's air were 0.21 parts per million, almost three times the limit. This was a common issue for most industrial areas in Michigan that were home to manufacturing facilities, which totaled twenty-one statewide in 1982.[20]

Problems on the Ground

Illegal dumping was also an issue. In one notorious case in 1978, residents in Gaines Township finally won a two-year battle with the Michigan Air Pollution Control Commission to stop the operation of a smoking incinerator owned by Berlin and Farro Liquid Incineration Inc. But soon thereafter, Gaines Township residents became "irate" when they found out that the company would be permitted to build an incinerator just thirty minutes away in Flint.[21] Berlin and Farro had a history, having been punished by the DEQ for illegally dumping industrial wastes into creeks and ponds, burying waste, and dumping liquids into agricultural drains since 1971 (and then continuing such activities, even after it briefly lost its permit to operate in 1975).[22]

Unfortunately, Berlin and Farro had ended up doing what residents feared most: polluting the air, contaminating the water, and endangering their lives. And these actions created one of the worst toxic waste sites in the nation. "Tens of thousands of cubic yards of solid waste and sludge and thousands of drums and buried tanks" were discovered on the company's property.[23] Nearly 170 residents had to be evacuated for up to a month while the federal EPA tried to deal with the area. One year after cleanup efforts—which cost almost $25 million between 1983 and 1996—the EPA released a report indicating that the site was associated with liver and kidney damage in humans as well as nervous system disorders.

Unfortunately, this report did not make it to Ron Voelker, a thirty-six-year-old construction worker who, in 2004, purchased a home on South Morrish

Road about a quarter mile from the Berlin and Farro toxic dump site. After living in his home for three and a half years, he developed an aggressive form of liver cancer and was told that he had six months to live. Voelker had no clue that he was purchasing a home on contaminated land for his family—which included his wife and five children.

"I never knew there was a toxic dump here," Voelker told the *Flint Journal*. "I feel like I put my family in harm's way."[24]

Almost eight months after his diagnosis, his eighteen-year-old daughter, Shyra, was diagnosed with Hodgkin's disease after doctors discovered a "grapefruit-sized tumor in her chest." Voelker passed away from liver cancer on March 27, 2009.[25] His daughter died almost six months later, on September 20.[26]

The previous owner of the property, Beth Agle—who purchased the home in 2000 with her husband and two daughters—was also diagnosed with ovarian cancer in 2003 (which then spread to her bladder). She and her family did not know that the water in their home was contaminated. As Agle told a *Flint Journal* reporter in 2008, "I wasn't aware of the site being there. We drank out of the well all the time, and we didn't think anything of it . . . and now it makes me wonder."[27]

Trash That Piles Up

While companies like Berlin and Farro were dumping toxic waste—and the apparatus of the State of Michigan was continuing its habit of granting them carte blanche—antipollution regulations actually did give private citizens a voice in the battle for a cleaner and safer environment. In Flint, block clubs—which had grown from 25 in 1952 to 2,000 in 1971—began applying pressure on city departments and industrial facilities to control pollution. Operating on the principle that "the squeaky wheel gets the grease," these block clubs teamed up with environmental groups like the Flint Environmental Action Team to appeal—persistently—to city officials for help in protecting their homes and neighborhoods from blight and pollution.[28]

Many citizens were concerned about landfill siting decisions and began advocating from a social justice angle, just as more research was emerging that linked health problems to residing near sites with radon exposure, acid rain, medical waste, and other hazards. During the 1980s, consistent pressure by community advocates closed at least three landfills in Flint, while five remained open.[29]

The city was less responsive about other concerns, such as the environmental hazards posed by inadequate rodent control and the overall conditions of local dumps. Arthur Hendrickson, a Flint air pollution specialist with thirteen years of experience, told the *Flint Journal* that he had had a front-row seat for the city's negligence regarding pollution and other citizen complaints. "People have had their dogs killed by rats from the dumps [in] the N. Dort Highway area," he recalled, "[and] doctors have told some residents to move for their health."

Beyond complaints about aggressive rats, smoke from dump fires at the Richfield Road junkyard and rancid black smoke from the Flint sewage treatment plant endangered locals, according to Hendrickson. Throughout the early 1970s the city ignored these issues in violation of its environmental ordinances and policies. As examples, Hendrickson shared that the city had been leaving dead animals and garbage in the city landfill at Bishop Airport—items that should have been under at least six inches of dirt. Hendrickson also mentioned dumping violations at Swartz Creek near Asylum Street, the east side of Groveland Avenue, and the Flint River.[30]

Illegal landfills had sprung up, too. Flint Park, a former amusement park located northwest of West Stewart Avenue and Dupont Street, had been on residents' radar since the early 1980s. It had become an eyesore because of the extensive accumulation of garbage, dead animals, appliances, and old furniture, all piled under a sign that read "No Dumping—Police Orders." After months of pleading with officials to block access to the park and deploy cleaning crews, residents finally got the city to agree to build metal barriers at Winthrop and West Flint Park boulevards. The restrictions, however, never truly deterred dumping; people wanted to sidestep tipping fees at local landfills. According to Clarine Patterson, president of the Flint Park Neighborhood Association, locals began advocating for stricter enforcement of the city's dumping laws. Neighborhood residents were soon reporting license plate numbers of people spotted illegally dumping.

"I figure getting license plate numbers should be enough," said Patterson. But the police seemed to "want the history of your life" when they reported offenders. "Why can't they put the pressure on the individuals who are doing the dumping?" The park's owners were little help—the State of Michigan owned a portion of the land, while the majority was held by the City of Flint.[31]

Things got worse when, in anticipation of revenue-sharing reductions due to Governor Milliken's property tax relief plan, Flint mayor James Rutherford announced a plan to temporarily replace weekly garbage pickup with biweekly pickup, from September 1981 to May 1982. When some of the city council members voiced opposition, the mayor insisted they were only kowtowing to voters in anticipation of that November's election.

Council members fired back that the city's financial situation was not actually shaky enough to require going to an every-other-week trash schedule. "It is my impression," said Second Ward councilman Floyd Clack at a special meeting, "that the people would like the weekly pickup to continue right now." The city council, however, had no power to override the mayor's decision, nor could members change the annual budget after they had approved it. All they could do at this point was voice their opposition.[32]

In raising the alarm, council members were joined by Fred M. Germaine, Genesee County's director of environmental health services. He pointed out that an experiment with biweekly garbage pickup the previous year had led to an explosion in the city's rat population. Despite the back-and-forth—and multiple pleas to delay the decision for the sake of Flint residents—Mayor Rutherford stood by his plan, noting in a letter to city council members, "We told the people that next spring's garbage would be picked up weekly at a time when rodents are more likely to manifest themselves. Gentlemen, that is what we agreed upon."[33]

Yet the snow melted as spring came and went. By the fall of 1982, Flint residents still did not have weekly trash pickup. And the city, as an impending $500,000 budget cut loomed, still didn't have the revenue to accommodate all of its needs or fulfill residents' desires. Council members and others fought over priorities, such as the importance of repairing potholes (and thereby limiting the damage to motorists' vehicles). City administrator Nan Lunn concluded, however, that garbage had to be a top priority since "pickups are the biggest complaint we get from people. They have been patient [with biweekly pickups]. They deserve it. They paid for it." Weekly pickups resumed the following Monday—in December, some seven months later than the city had promised.[34]

In some cases, during these tough financial times, residents in low-resource areas took it upon themselves to clean up their neighborhoods. Civic-minded folks like Inella Biggs, who lived on East Philadelphia Boulevard, took the self-help approach by organizing a garbage pickup in her neighborhood. Yet the city refused to donate garbage bags and did not allow employees to pick up the trash residents collected.

Without labor from federally subsidized programs like the Comprehensive Employment and Training Act, trash was not collected on vacant lots. So residents would collect and bag it, but no one would then retrieve the bags. The result would be seventy-five bags of garbage sitting and stewing in the languid Michigan summer, stacked on the curb of East Philadelphia Boulevard for over a month. Technically, there was no limit to the number of bags residents could pile on their curb, so long as the bags were intact, weighed less than fifty-five pounds, and were positioned within five feet of the curbside. But, as a city maintenance employee noted, the city did not have the capacity to deal with this volunteer effort; Flint could barely keep enough people employed to cover regular trash pickups.

"It's admirable that they would want to clean up the city, but they should

check first with somebody [from the city] if they're not going to dump it themselves," this maintenance employee said, adding that his coworkers "do a lot of
things over and above what is required, but if they see 200 bags stacked up in
front of one house, no way."[35]

Gasoline: A Lurking Danger

During the mid-1980s, Flint residents had something else to add to their growing list of environmental complaints: water pollution from leaking and abandoned underground gasoline tanks that were being discovered with astonish-
ing frequency.

"It's getting way out of proportion," said a local DNR specialist, Ben Hall.
"Calls are coming daily now, where we get one or two tank removals; some are
just upgrading their tanks, but others have a problem." Genesee County began
installing monitoring wells and digging up soil, concentrating on corner gas
stations and the other sites with abandoned tanks, in order to measure the pollution levels of properties long overlooked not only by their owners but also by
the DNR.[36]

Although the public's knowledge about the medical dangers of lead contamination was limited, people had become fearful of it leeching into their surroundings. By the time lawmakers began pushing for the removal of lead from
gas in 1977, an alarmingly high amount of it had already been discovered in
Michigan urban areas. Lawmakers in Washington, DC—including Michigan's
own Senator Philip A. Hart—contended that lead-saturated soil was especially
problematic for children and put them at risk of ongoing health problems, including mental retardation if they ate or sucked their thumbs without washing
their hands. Soil was just that contaminated in many industrial cities.

One of the most significant findings from an unpublished EPA report was
the overconcentration of lead-contaminated soil in residential areas. In 1972,
within nine Michigan cities, the EPA found an average lead content of 1,339
parts per million (ppm) in business districts—but 2,372 ppm in residential
areas (in Flint, the numbers were 2,689 ppm and 3,867 ppm, respectively).[37]
As table 2.1 illustrates, in some of the communities included in this study the
proportion was flipped; samples from Ann Arbor found averages of 1,871 ppm
in residential areas but 4,269 in commercial areas.

Lawmakers were stymied. Many assumed that the lead problem had been
properly traced to automobile exhaust from burning leaded gas. Senator Hart
wrote to the EPA, urging the agency to strengthen gasoline regulations: "We
conclude that neither cost nor competitive considerations provide a sufficient
reason for abandoning a schedule of total lead removal by 1977[, because] . . .

Table 2.1. Lead-contaminated soil in residential and commercial areas in 1972 (as expressed by parts per million)

City	Residential	Business	Percentage Black
Ann Arbor	1,871	4,289	7
Grand Rapids	2,138	3,663	7
Jackson	2,206	2,524	8
Kalamazoo	2,851	4,722	5
Lansing	1,941	2,637	4
Pontiac	2,761	2,780	27
Saginaw	3,042	2,527	18

Sources: Campbell Gibson and Kay Jung, *Historical Census Statistics on Population Totals by Race, 1790 to 1990, and by Hispanic Origin, 1970 to 1990, for Large Cities and Other Urban Places in The United States*, Population Division, Working Paper No. 76 (Washington, DC: US Census Bureau, 2005), Table 23: Michigan—Race and Hispanic Origin for Selected Large Cities and Other Places: Earliest Census to 1990; 1970 Census of Population Characteristics of the Population, Michigan, table 23, 77–79; Robert Lewis, "Study Indicates High Lead Level in Flint's Soil," *Flint Journal*, May 27, 1972.

if we do not adopt, this schedule . . . we will continue to expose the public, and particularly the inner-city poor, to substantial hazards."[38]

In 1986, against the protest of service station lobbyists—who claimed new regulations would put them out of business—the federal government unveiled a new program to clean up lead problems that stemmed from leaking underground storage tanks (boasting the headline-grabbing acronym LUST). Before 1986, few such tank violations were reported, but thereafter the DNR could barely keep up with violation reports. Some 35 percent of the nation's underground storage tanks were estimated to pose a threat to soil and groundwater, with the problem compounded in large industrial areas. A city devoted to the auto industry, Flint would see the removal of nearly 400 underground tanks in 1986, 300 in 1987, and 150 by August 1988.

Gas station owners, especially mom-and-pop operations, were unlikely to report problems. The average cleanup cost was $100,000, and few could afford that. Instead of reporting known leaks, these businesses generally just closed if spills were discovered. For example, in 1989 Ed Bathish, owner of Ed's Eastside Mobil station at East Court Street and Center Road in Flint, discovered that a pipe connected to a storage tank under his gas station was cracked and leaking gas. He mistakenly thought that the 0.875 percent gasoline tax he paid to the

Michigan Underground Storage Tank Financial Assistance Fund would cover
the spill, but his funding request was denied because he had missed the deadline to submit a report.

Bathish and his wife had struggled to build their business over a span of
twenty years and were then forced to close their doors in the face of the $150,000
estimate to clean up their property after the contractor who had begun the work
sued them. Other gas stations had similar stories. In 1991, Action Auto in Burton was required to clean up the gasoline and benzene discovered in groundwater that had been caused by a leaking 11,000-gallon underground storage tank
that had been removed almost six years earlier. The state compelled the owners
to do a complete assessment of the contamination[39] and to pay for the cleanup
of the groundwater and soil, plus an additional $10,000 in civil fees.[40]

In 1989, eighteen residents on North Term Street in Flint, including Michelle
Heiser, complained that their well water reeked of gas and made their families
sick. The Heiser family had lived in their Genesee Township home north of
Richfield Road for six years before encountering any problem with their well
water. They initially attempted to ignore the issue, using a lot of Kool-Aid and
Tang to try to hide the smell. The city eventually tested their water and discovered it was not safe, with benzene levels—astonishingly high, in fact. In thirty-
foot wells near North Term Street, benzene levels were 250 times the "safe"
level. It was the worst case of underground drinking water contamination in
Genesee County. Residents became dependent on bottled water.[41]

Once it was confirmed that wells were contaminated, the state in 1990 approved a $710,000 project to seal them off and connect almost 140 homes to
the city's water system. Residents on neighboring streets were connected when
their well water was also discovered to be contaminated with gas. Citizens on
North Term were glad to hear that the state was doing the right thing by protecting them from tainted well water, but most remained cautious about the
government's intentions, vowing to avoid drinking tap water until the project
had sealed off the wells entirely. Heiser reported that she still did not plan to
drink the water: "They say it (city water) is safe to drink. [But] they've been
telling me several years that my well water is safe to drink." Getting the city and
state to act had taken years—so would rebuilding trust.[42]

Opposition to Environmental Advocacy

During John Engler's tenure as governor in the early to mid-1990s, Michigan's
pollution rates peaked faster than those of any other state in the nation. At the
same time, environmentalists and conservationists began to have fewer opportunities to voice their concerns about protecting natural resources.

In 1991, Engler abolished the DNR that had been created by the Michigan legislature and reorganized it into a new DNR that was headed by the chair of the Natural Resources Commission—who was appointed by the governor. By the end of 1993, legal challenges by the Michigan United Conservation Clubs and the Michigan Environmental Protection Foundation to Engler's abolishment of the legislatively created DNR were unsuccessful in forcing him to restore the independence of the Natural Resources Commission or to disband the Michigan Environmental Science Board he created with Executive Order 1992–19.[43] Instead Engler made additional attempts to reconfigure the state's regulatory infrastructure for pollution control with measures designed to undermine its accountability to citizen review, which included his move to eliminate oversight panels charged with conducting public hearings regarding water and air pollution permits. Under Engler's plan, the authority to issue pollution permits was controlled exclusively by state bureaucrats within the DNR, the Natural Resources Commission, and the governor's ofice.[44]

Engler tried to further reduce the size of state government by declining to fill vacant positions in state agencies, including the DEQ. His administration restructured the way decisions about environmental permits and sanctions would be made. Employees claimed their biggest hurdles in enforcing environmental statutes were "insufficient staff to do appropriate number of inspections," "an anti-enforcement mentality within the Executive Office," and a "pattern of substituting political agendas for professional scientific evaluation."[45]

Morale within the DEQ suffered. A survey conducted by the Public Employees for Environmental Responsibility in 1998 captured the concerns of the 41 percent of Michigan's DEQ employees who responded. More than half of them feared retaliation if they enforced environmental laws and noted that employees had been reassigned or transferred for "doing their job 'too well' on a controversial project."

Over a quarter of the responding employees reported having been instructed to disregard environmental policies. They seemed to agree that the chain of command interfered with their capacity to enforce environmental protection laws. According to 83 percent of the DEQ respondents, economic development was more important to the administration than public or environmental protection. The pro-business Engler was, they alleged, allowing corporations to dictate enforcement actions to suit their economic needs (70 percent contended that industry and developers "excessively influence[d]" DEQ rulings). One employee noted that "the regulated community is able to schedule meetings on short notice with the governor, director and/or division chief. District staff involved with the issue are not invited to attend and are not asked for briefings. The regulated community always gets what they want, at the expense of the environment and public health, while the DEQ staff morale and trust goes down the toilet."[46]

Patterns of pollution and lax enforcement that were emerging supported the concerns that were being expressed. For instance, after Engler took office, DEQ environmental enforcement decreased substantially: 149 new cases had been opened in 1990, but only 76 in 1998. By July 2000, according to the Environmental Working Group, Michigan ranked as the fifth worst state when it came to inspecting known violators of laws relating to air quality and the third worst regarding notorious violators of water laws. In 1997, the Engler administration undermined citizen oversight by vetoing a $250,000 legislative appropriation for right-to-know programs. During this time, the DEQ's authority was undermined and businesses were allowed to overlook their obligations to report use and discharge of toxic materials—as was stipulated in state law.[47]

It would take intense public pressure to force Engler's administration to accept a $4.9 million federal grant to protect Michigan children from lead poisoning. The governor's office had issues with the federal demand that the licensing requirements of lead remediation contractors be regulated. The administration's inaction delayed the state's efforts to address childhood blood screening for lead (the state's rates were among the worst in the nation). Only 7 percent of an estimated 40,000 children under age six who lived in poverty and had an elevated risk for environmental lead exposure were screened annually.[48]

As DEQ employees voiced concerns with Engler, residents throughout Michigan were stuck living in under-regulated spaces that threatened their health. Just after Engler reorganized the DEQ in 1995, giving the agency responsibility for regulating health and sanitary conditions in trailer parks, the City of Ecorse requested help with the Hannah Park Trailer Court. The area was overrun with rats, roaches, and garbage and contained many illegal natural gas and sewage hookups. Circumstances escalated when a fire broke out in a home that was constructed with two trailers illegally joined together. The occupants almost lost their lives. Still, the DEQ took no action against the owner of the trailer park. It conducted the legally mandated annual inspections at the time of the initial complaint by Ecorse but failed to note gross health and safety violations. Nor did it do so three years later. In 1998, according to the DEQ, Hannah Park was in "substantial noncompliance."[49]

By February 26, 2000, public outcry and subsequent news reporting describing children playing in raw sewage during snow melts compelled Ecorse's fire chief to invoke his emergency powers. The chief condemned the trailer park—citing cat-sized rats, dirty syringes littered about, rigged electrical and gas lines, and raw sewage.

Engler vs. Flint's Environmental Justice Advocates

Genesee Power Station

Engler spent a great deal of time and energy during his gubernatorial tenure attacking what he called the EPA's "reckless, ill-defined policy on environmental justice."[50] The news media, especially David Mastio of the *Detroit News*, unabashedly supported him.[51] Mastio published several articles in which he consistently admonished the federal agency for increasing its scrutiny of businesses in poor communities that were in need of economic development. According to Mastio, attempts to link environmental pollution with racism in poor communities were misguided and dangerous to mid-Michigan areas that badly needed economic investment. His belief was that vague efforts to protect poor racial minorities from environmental hazards were smothering attempts to secure them jobs. Mastio cited comments from Michigan congressmen, such as US representative Joseph Knollenberg of Bloomfield Hills (who characterized the EPA as "intellectually bankrupt") and US representative James Barcia of Bay City (who claimed the EPA's actions were not supportive of economic development in Michigan).[52]

During this period, on June 8, 1992, the Genesee Power Station Limited Partnership reached out to the Michigan Air Pollution Control Commission to apply for a permit for a 35-megawatt power plant using wood waste as fuel. It was to be located on the north side of Carpenter Road in Flint, within a three-mile radius of fourteen schools. Residents initially learned that an incinerator or wood-burning power plant was being considered when the *Flint Journal* ran a story about the project in 1991. They turned to Genesee Township supervisor William C. Ayre, Flint city councilman Jonny Tucker, and Genesee County commissioner Vera Rison to learn more. Ayre promised that he would inquire into the matter but never followed up. In fact, no further information about the incinerator emerged until one of the petitioners overheard a Michigan Public Services Commission employee noting, in September 1992, that Genesee Township had recently approved a permit for a wood-burning plant. The DNR issued a public notice for the proposed facility, released a draft permit, and set a public comment period that would end in a matter of weeks, on November 17, 1992.[53]

Organizations representing the public—including the American Lung Association of Michigan, the Flint chapter of the NAACP, the Genesee County Medical Society, the Flint Neighborhood Coalition, and the St. Francis Prayer Center—fought the incinerator permit throughout the year. They added their pleas at a public hearing on October 27, 1992. And yet, the DNR elected to move forward, granting the permit in early December 1992, before its final draft or associated staff reports concerning its environmental effects were even available to the public.

Residents were unable to submit detailed written comments on the final draft and were dismayed that it included none of the stipulations they had so vigorously advocated for. These included a prohibition against adding painted materials to the incinerator and a requirement that it use emission control technology. In addition, they wanted better monitoring of toxic emissions that were being emitted than what the Best Available Control Technology could provide (which residents contended was inaccurate in detecting problems with fuel content).[54]

Locals chartered a bus to the state capital in order to express their concerns about the proposed facility at a second public hearing on December 1, 1992. The Michigan Air Pollution Control Commission meeting began in the morning; commissioners reviewed the final permit draft and relevant staff reports and distributed them to the public at about two o'clock. Residents were not allowed to offer comments until seven that evening,[55] and even then, their concerns fell on deaf ears. The DNR's Air Quality Control Commission (now known as the Air Quality Division) determined, by a vote of 6–1, that Genesee Power Station's permit met all federal and state requirements. This commission issued a notice on December 7 indicating their intent to approve the permit, effective January 9, 1993.[56] Several residents requested an additional comment period but were denied.[57] The State of Michigan did not perform risk assessments described as "normally desirable,"[58] nor did the DNR undertake epidemiological studies to ascertain lead exposure rates or how the presence of a wood-burning power plant would contribute to such rates.[59]

The population was already contending with a host of environmental hazards. On Flint's north side, junkyards were burning tires and trash. Gas storage tanks were leaking, a fenced-off holding pond on the northwest of the proposed site was filled with liquid waste and sludge, and the Flint River—already the second-most polluted in Michigan—flowed through both the proposed site and nearby residential areas. Though the DNR was of no help, nine community organizations fought the project vigorously. These groups eventually sought help from the EPA Appeals Board, which decided on October 22, 1993, that the DNR must reconsider its decision to use the Best Available Control Technology to monitor lead emissions and also required Genesee Power Station to not burn wood that was coated or treated with lead-bearing substances.[60] The DNR, however, never required Genesee Power Station to comply with the terms of this order.

Opponents of the power station took the state and responsible state agencies to court. The NAACP and the St. Francis Prayer Center spearheaded the

Figure 2.2. Junkyards eliciting resident complaints, March 18, 1993. (Genesee
Power Station, St. Francis Prayer Center Records, Bentley Historical Library,
University of Michigan)

complaint, alleging that the DNR's decision to issue the power station a permit violated state human rights regulations. By this time, the Governor's Science Board Lead Panel had released a 1995 report, "Impacts of Lead in Michigan," that noted that nearly 50 percent of Flint's children between the ages of six months and five years had elevated lead levels in their blood (above 10 micrograms per deciliter, the threshold set by the Centers for Disease Control at that time). Flint's percentage of childhood lead exposure ranked third on a list of thirteen cities in the state, just behind Detroit and Battle Creek (which had childhood lead exposure rates of 57 and 51 percent, respectively). Flint's elevated levels were concentrated on the north side, where the Genesee Power Station would be sited. Residents again raised their concerns, fearing for the safety of some 60,000 people living within a three-mile radius of the proposed station.

Both before and after the lawsuit, the DNR contended that the only relevant information in its decision involved data relating to potential *air* regulation violations. That meant that the DNR could dismiss most of the commentary from the public—whether residents, scientists, physicians, or even air pollution experts—as irrelevant. It refused to consider testimony from scientists who claimed that the plant would contribute to the process of poor and minority communities being targeted for undesirable land uses and also rejected a petition with over 1,000 signatures as being beyond their scope of review. Children in this area already faced an elevated risk of lead exposure, but the DNR was pushing forward with a power station that would be placed directly across the street from an elementary school.

The judge in the case urged the parties to settle. While the plaintiffs and the Genesee Power Station were able to agree that the facility would limit the burning of construction and demolition wood to 20 percent while also being subjected to the supervision of a "Special Master," the plaintiffs and the DNR were not able to come to a final settlement. The case went to trial in the spring of 1997.

The DNR's primary defense was that it did not have the authority to deny permits based on degrees of environmental risk or demographic factors. It could consider only whether the permit seeker's emissions were within National Ambient Air Quality Standards. On May 29, 1997, the court rejected this defense, stating that in issuing the permit, the DNR had neglected to protect the health and safety of Flint residents. The court dismissed the plaintiffs' contention that the permit review process had a disproportionate impact based on race but approved the plaintiffs' motion for a permanent injunction. The decision asserted that the disproportionate siting of waste facilities in Flint was the result of flawed state policy.

In the meantime, the Genesee Power Station was issuing visible pollution (see figure 2.3), and the public controversy roiled Flint. On July 7, 1994, the

Figure 2.3. Genesee Power Station at morning and night, ca. 1994. (Genesee Power Station, Miscellaneous 77, St. Francis Prayer Center Records, Bentley Historical Center, University of Michigan)

Sugar Law Center for Social and Economic Justice, which represented Flint residents protesting the permit, appealed to the state DNR through the federal EPA. The state agency missed its decision deadline by almost fifty days but rejected the complaint on September 16, 1994, as "untimely." But this decision would be reversed when the EPA discovered the *original* complaint that had been filed by St. Francis Prayer Center on December 15, 1992. Nearly two years after the Sugar Law Center's Title VI of the 1964 Civil Rights Act complaint had been filed, it was finally accepted by the EPA on January 31, 1995. But then the EPA dragged its feet and, instead of replying to the complaint within the prescribed thirty-day period, it accepted the DNR's response on June 26, 1995.[61]

It is worth mentioning that the EPA's backlog was not specific to Flint. The agency perpetually delayed acknowledging, investigating, and adjudicating Title VI complaints. For instance, after African Americans for Environmental Justice filed a complaint in September 1993, the respondent—the Mississippi DEQ—ignored the complaint. Some 150 days after the deadline, the EPA asserted that it must respond by April 15, 1994. Yet another year would pass until it actually did so—on April 11, 1995. And it would take until August of that year before EPA officials visited the site and met with the complainant. The wait for investigations and adjudication was so long that complainants from across the United States joined together to protest the sluggish process.[62]

The EPA claimed that delays in processing Title VI complaints were driven by the need to establish its Interim Guidance for Investigating Title VI Administrative Complaints Challenging Permits, to improve its methodology, and to await methodological recommendations from its science board.[63] It turns out that the EPA's process of perfecting its methodology would take longer than anyone could ever have imagined. Despite receiving the initial Genesee Power Station complaint from Flint residents in 1992, this complaint was not resolved by the EPA until January 2017. That's a quarter of a century.[64]

Throughout the back-and-forth regarding the Genesee Power Station's permit application, it was clear that the DNR was unaccustomed to considering permits' disproportionate impact on poor communities of color. In Kent County, a municipal solid waste incinerator was placed in an area dominated by Black residents (even though Black people made up only 8.1 percent of the area's total population). In Macomb County—just over 1 percent Black in total—an incinerator was situated in an area where the Black population was nearly 10 percent. In Wayne County, a pair of municipal solid waste incinerators were placed in predominantly Black areas (54 and 80 percent, respectively).[65] When asked about the DEQ's efforts to prepare guidelines on how to act on environmental justice concerns, director Russell Harding said that the agency had made no effort to protect against environmental racism. Additionally, it had no plans to hire a specialist to evaluate the racial effects of its decision-making process.[66]

Select Steel

Just after the trial court's May 1997 decision to dismiss the claim of racial discrimination in the granting of a permit to the Genesee Power Station, the DNR obtained a stay of the injunction and issued a permit to the Dunn Industrial

Group to locate a "mini" steel mill, dubbed the Select Steel Mill, in Flint.[67] Characterizing the EPA as a "job killer" that heaped "unsupported burdens on businesses," Governor Engler began a publicity blitz calling for the EPA to "get out of the way" of projects like Select Steel. The proposed steel mill was a $175 million project expected to bring hundreds of jobs to the area, and the permit application was submitted on December 30, 1997.[68] A few months later, Dan O'Brien from the DEQ Toxics Unit sent an interoffice memo to Hien Nguyen in the Permit Unit that expressed concerns about Select Steel's mercury emissions. O'Brien noted the toxicity and bioaccumulative properties of mercury, arguing, "If the facility were allowed to operate at this emission limit, annual mercury emissions would be among the highest in the state." He claimed that his colleagues agreed that the DEQ must address concerns about mercury emissions before issuing the permit.[69]

This was not the Toxic Unit staff's last interaction with the rest of the DEQ and the EPA regarding Select Steel. The unit had responded to various requests for data about the complaint. But the DEQ's deputy director, W. Charles McIntosh, noted in correspondence to the EPA's Office of Civil Right (OCR) dated September 30, 1998, that he was denying further access to DEQ staff and records with regard to the Select Steel and Genesee Power Station projects, though the Toxic Unit would produce all documents requested by the EPA and its Office of Civil Rights *except for* documents regarding mercury. McIntosh appeared clueless about why the EPA and the OCR would be interested in such data. According to him, "since mercury was not mentioned as an area of concern in the complaint filed against the DEQ, please explain why this information has any relevance to the investigation of the DEQ's issuance of a permit to ... Select Steel Corporation of America." The DEQ claimed that it was not its intention to "halt or derail these investigations in any way," but it appeared the agency knowingly neglected mercury concerns in order to fast-track the Select Steel project in Flint.[70]

To counter arguments that the Select Steel project disproportionately affected racial minorities in the area, the *Detroit News* commissioned a report by Patrick Anderson of Anderson Economic Group. It claimed that assumptions about the demographic composition of the area were false and argued that the one-mile area around the proposed site did, in fact, have white residents. But the report mischaracterized the area, reporting the racial averages of four census tracts surrounding the proposed site (84 percent white) to dilute the reality of the census tract that directly bordered the site (which was 71 percent Black).[71]

Even so, Engler and the *Detroit News* failed to intimidate vocal activists. The St. Francis Prayer Center reached out to the EPA, again requesting that the agency consider its environmental justice concerns to overturn the DEQ's decision to grant Select Steel an air quality permit. In 1998, Flint residents also filed an administrative complaint with the OCR against the DNR in which it was alleged that the department failed to consider the disproportionate burden of pollution on predominantly minority communities. Residents went on to claim that the Select Steel decision violated Title VI of the 1964 Civil Rights Act and that Select Steel did not adequately entertain public comments about the operation of a forty-three-ton-per-hour steel mill in their community. Petitioners contended that one public hearing and two informational meetings were held in white schools, while neither hearings nor meetings were ever held in the predominantly Black neighborhoods that were voicing concerns with the permitting process.

In response, the State of Michigan maintained that there had been a thirty-three-day public comment period and a public hearing (held April 28, 1998) that eighty-two people attended, with twenty-nine providing public comments and thirty-five submitting written ones from businesses and residents (thirty-one in support of the steel mill and four opposed).[72] The state also maintained that it had made serious efforts to assess the likelihood of children's lead exposure before approving the permit on May 27, 1998.[73] The DEQ concluded that "the negligible potential impact on blood lead levels did not merit denial of the permit application."[74]

Governor Engler was personally critical of the EPA complaint filed by Reverend Phil Schmitter and Sister Joanne Chiverini, codirectors of the St. Francis Prayer Center. But Engler's beef was *really* with President Bill Clinton's Executive Order 12898, which required federal agencies to collect data on health and environmental risks and help prohibit the "disproportionate effects" of discrimination based on race. Engler believed this order altered the traditional notion of racial injustice—defined as overt actions to deny civil rights based on race—to focus on disproportionate or discriminatory effects of actions by state agencies that received EPA funding.[75]

The St. Francis Prayer Center claimed that Select Steel's permit was deficient because it included no monitoring requirements for lead or dioxin, while the DEQ claimed there simply wasn't any available technology to allow for such continuous monitoring. The complainants challenged the DEQ's decision to give Select Steel up to two years to meet volatile organic compound monitoring requirements, while the DEQ argued that volatile organic compound emissions monitoring was not required under federal law. The EPA Appeals Board denied a review of the St. Francis Prayer Center's petition on September 10, 1998, arguing that it had neglected to specify and prove the effects of the DEQ's decision-making process or why the agency's decision to issue the permit was problematic.[76]

During this time, the *Detroit News* began publishing stories criticizing the EPA's decision to dismiss the administrative complaint out of hand. Select Steel's owners were, at this point, threatening to move the plant (and its jobs) to Toledo, Ohio, if the EPA investigation lasted more than forty-five days. Governor Engler intervened by staging a press conference in Flint and demanding that the residents withdraw their complaint. They continued to resist but ultimately did not triumph. In its first formal decision, the EPA's OCR ruled against Flint residents, concluding that demonstrating a business was a potential polluter was not enough to prove a disproportionate or discriminatory impact on a community by a siting decision. The petitioners were told that they would need to present data illustrating that the emission levels would rise above legal levels to prove that the decision posed harm to the environment and disproportionately affected vulnerable communities. On October 30, 1998, the EPA found that the DNR had not violated the civil rights of Black residents in granting the permit. Though the mill would discharge various pollutants, including mercury, the EPA concluded that there was no *disparate* adverse impact because emission levels would not exceed current environmental standards.[77] The mill could move forward.

In the end, Select Steel opted to forgo the Flint site in favor of one in Lansing. But the St. Francis Prayer Center did not let up, arguing that the EPA's decision hindered efforts to prove environmental racism regarding siting decisions in poor minority communities. Julie Hurwitz of the Sugar Law Center wrote that the Select Steel decision set a bad precedent: the EPA had issued a decision that was "not based on sound evidence or analysis" but was influenced by "unrelenting political pressure from right-wing advocates and from Michigan decision-makers."[78]

In a rhetorical turn that would foreshadow Trump-era anti-regulation theatrics, Engler and other supporters of the Select Steel project contended that "stopping this urban regulatory death knell can't happen soon enough."[79] They asserted that concerns about environmental racism were "neither justified nor scientific" and argued that the EPA had not been able to produce any study linking health issues with pollution problems. Other proponents, including Detroit mayor Dennis Archer, asserted that racial disparities in health outcomes were in fact shaped by the actions of poor minorities who opted to move to cheaper areas in cities where old factories and waste sites were located. In fact, in the DEQ's response to the St. Francis Prayer Center's complaints, the state agency had argued that "unlike an employer applying its facially neutral policy to its employees, the M[ichigan] DEQ did not choose the site of Select and does not control it. Select Steel chooses where to build in connection with local zoning authorities. Additionally, the MDEQ does not control where people choose to live." The department abdicated responsibility, saying that state agencies did not possess the right to decide for local communities, or their elected officials, whether a business should locate in a specific site.[80]

The document went on to claim that the EPA and DEQ had no expertise in determining where plants should be sited; rather, "local communities are the best able to weigh the benefits and determents of a particular industry locating in their communities and to decide whether they want it to."[81] The DEQ further expressed shock that the St. Francis Prayer Center's Title VI complaint could be raised without evidence of intentional discrimination. Such activities, the DEQ warned, would only, in the end, hurt poor communities. "Regrettably," the DEQ contended, "the well-intentioned policy of applying a disparate impact model to the granting of an environmental permit may only exacerbate and prolong existing unemployment, health problems, and lack of educational opportunities for minorities by discouraging industry from locating in depressed urban areas with significant minority populations."[82]

If conflicts around the Genesee Power Station and Select Steel hadn't emerged in Flint, the DEQ's argument about community power in siting decisions might have held water. For years, Flint residents like Lillian Robinson of United for Action, Justice, and Environmental Safety worked with neighborhood residents to deal with the consequences of small-source pollution from business establishments, including auto body and paint shops, in mixed-zoned areas. Robinson had lived on West Boulevard Drive in Flint for over four decades. In her early seventies, she was a strong advocate for justice and disturbed by the "things that's happening against the people that can least help themselves." She was particularly concerned about what the effects of living in a highly polluted area might be on young single mothers and their babies, children attending the neighborhood school, and those living in two large apartment complexes. As she put it, "To take a child, tell them you must go to school, you must learn, you must try to do something with your life, and then put polluters all around them to destroy what they have to use to learn with, and that's their brain. Lead poison, arsenic. All the things that come from incinerators [and] steel mills."[83]

For a while these activists thought they were getting through to the Air Quality Control Commission. Still, Robinson charged, Governor Engler "decided he would fire the Air Quality Control Commission that was listening to us and saw that you know, this was a real problem; he fired them." Her group then appealed to the DEQ, but the "DEQ was more for business than people. They have totally ignored us."[84]

Their next stop was the federal EPA, which Robinson said was clearly "under pressure from political machinery all the way from the White House to the state."[85] Ultimately, the community that the DEQ claimed had the rightful power to decide whether to prevent or pave the way for new industrial sites was robbed of that power at every turn—including by the DEQ.

Meanwhile, civil rights–focused environmentalists, who fought for environmental justice, continued to intensely critique Engler. However, instead of cleaning up the spills from Michigan's 8,000 leaking underground gasoline tanks, for instance, Governor Engler planned to let polluted water lie in *some* areas or cities. The administration said pollution would be tolerated in areas where cleanup would delay or discourage urban development.

Critics argued that Engler's willingness to ignore pollution in urban areas where primarily poor and minority people lived was irrefutable evidence of environmental racism. For his part, Engler contended that he wished only to find a more efficient and cheaper solution for dealing with polluted sites; the fear generated by expensive cleanups was terrible for business, and no one could risk that in an era of rising unemployment and factory closures.

One of Engler's advisors, Michigan State University biologist William Cooper, noted that the science did not support "cosmetic cleanups with no benefits." William Rustem, a Lansing environmental consultant who worked to restructure Michigan's environmental strategy, said that it just made sense to limit funding for environmental justice issues: "I would love to have every urban area pristine again, but there's not enough money to do that."[86]

The inauguration of a Democratic governor, Jennifer Granholm, in 2003 changed little. With regard to environmental injustices across Michigan, most of Granholm's attention between 2003 and 2010 went to pushing "smart growth" initiatives that involved significant investments in Michigan's roads, attempts to control urban sprawl, and making the state more energy efficient. Plans to mitigate environmental disparities developed toward the end of her administration but did not go far enough. This was especially true in regard to increasing public participation in siting decisions and providing a formal mechanism for transparent communication between regulatory agencies and Michigan residents.

For instance, on November 21, 2007, Executive Directive 2007–23 charged the DEQ with developing and implementing environmental justice policy for the state.[87] The resulting Environmental Justice Plan came out on December 17, 2010, after the DEQ merged with the DNR earlier in the year to create the Department of Natural Resources and Environment (DNRE) and just a month before Granholm left office. It was drafted by the Environmental Justice Working Group, comprising environmental justice advocacy groups, social scientists, tribal leaders, state agencies, and representatives from the business community. This group had met over the course of two years to develop guidance and recommendations for achieving environmental justice and "balancing productive economic growth with the high quality of life that is important to all people."[88]

According to the Environmental Justice Plan, the DNRE needed to develop and implement expectations and initiatives that were responsive to environmental justice principles, including enhancements to procedures for public involvement concerning projects proposed to be within a mile of minority communities and an agency process to identify and resolve disparate environmental impacts. The working group also identified ways of enlisting local government and expanding funding sources to support environmental justice plans. Michigan had, for example, received nearly $224 million from the federal government in 2010 to redevelop core urban areas in ninety-three census tracts in twelve major cities, including Flint, Detroit, Lansing, Benton Harbor, Grand Rapids, Highland Park, and Kalamazoo. Accordingly, $267,594 was available from the EPA's 2010 Pollution Prevention Grant Program in southeast Michigan to improve environmental outcomes and reduce energy consumption.[89]

Unfortunately, the *operational* policy enhancements the Environmental Justice Working Group envisioned to improve upon inspection rates, enforcement norms, and incentive programs never materialized. The plan was not implemented by the incoming administration of Rick Snyder. Department resources were not leveraged, and interagency activities were never restructured to maximize results. Governor Snyder quickly discarded the environmental justice recommendations and, in his first Executive Order, began another reorganization of the DNRE. "Reorganize" is a bit of a hedge, as Snyder simply moved to abolish the DNRE and reestablish an independent DEQ and DNR under the original organizational goals conceived by the Engler administration.[90]

The "Strange Fruit" of Benign Neglect

Unregulated disorder and lax regulatory enforcement in Flint's communities continued. Residents and city council members, in more recent years, continue to consistently complain about how the quality of city services and inadequate resolution to service reductions affected their lives and contributed to environmental hazards. Councilman Wantwaz Davis noted at a city council meeting in 2014, "We have burnt up houses with dead dogs in these basements, that bothers me because every time a house is burnt up, and you spray water on it, the ground is contaminated. Our children are playing in these dirty basements, and they're contracting something and passing it on to other children; now we have an epidemic of diseases going through the community, that is something we have to be mindful of when it comes to community development."[91]

The signs of benign neglect do not just manifest as broken streetlights and pockmarked roads, nor are they limited to the disgust of community members who are forced to sit down and shut up about their concerns. Benign neglect impacts the nature of "living free," whether from crime or environmental hazards or structural decay. The cumulative effects of such neglect were driving people out of Flint. One resident, Aaron Dionne, who was president of the Genesee Landlords' Association, stood up at a 2014 city council meeting to share that, without public protection, his neighbors (and renters) were trying to move away to more affluent parts of the county:

> Just this past week, Friday, a tenant of mine inquired about maybe moving to Flint Township because some property owner had moved out of the property next to him and there's a new family there that's dealing drugs. I said, "Well, call 911." "Well, they don't come," he says. And, I think I believe that. He says day and night, all hours, their customers are coming up, honking the horn in the road. I can't believe that would happen in normal conditions, but I guess that they're so brazen, people don't get in trouble for so many things, they don't worry about the police catching them.

Former city councilwoman Vicki Van Buren repeated this sentiment in another council meeting, referring to the effects of cuts in police and fire spending:

> Now, we're talking about cutting police and fire, then what? Do we close the doors on the city? I am not going to give up on a city that has done so much in its history and with its people. Right now, even in my ward, we're starting to see home invasions happening when people are in their house; either [morning or] night [they] wake up and see someone in their hallway. How bold are some of these criminals that think that they can run the city? What do we have to do to protect our families? Or are we trying to chase our families out?

Josh Billings, a local citizen, reminded residents at a January 2014 council meeting that the consequences of decades of benign neglect can be profound, extending well beyond criminal attacks:

> I've lived in this city over 40 years. I've seen way more snowstorms than what we got this last couple of weeks. The city has never been shut down a week. Our streets have never [taken] a week to be cleaned. My neighbor was cleaning the street out, not his driveway[,] when he died. [He was trying to] back outta his driveway, same problem I'm having. We got water running from a water break. It builds a hill, and the hill freezes and turns

into ice. So when you go down, you doin' one of two things when you're
going in and out your driveway. You either tearin' up the back end of your
car or the front end of your car. I submitted a complaint 'bout that, but it
still got ice and stuff like that. So, I say this: if we got our streets cleaned,
we might have prevented that death. That street [needed] to be cleaned
from the street to the driveway. His driveway was clean.[92]

Social disgrace and destructive behavior have long been blamed for wreck-
ing poor communities, especially after the emergence of the crack cocaine
epidemic in the early 1980s. Indeed, it seems self-evident—when streets are
littered with dirty needles or floods of people are coming in and out of drug
houses sprinkled across distressed communities—that drug abuse causes and
perpetuates disorder.

But as we saw in this chapter, there were plenty of Flint residents working
to do more than chase after habits and mourn dreams deferred during this period of the city's history. Most Flint residents continued the work of raising
their families amid declining federal and state support while simultaneously
adjusting to a severely declining job market—all while trying to protect their
neighborhoods, their homes.

The fact is that problems in poor Black areas throughout Flint intensified
when, during the mid- to late 1970s, city officials changed their approach to
managing discontent in these communities. They had already been undermined for decades by the local power elite's white supremacist agenda. At this
moment in the city's trajectory, attention and priorities shifted away from addressing problems in neglected Black areas—the primary recipients of civil
rights–era programs—to enhancing conditions in "historic" and more mon-
etarily valuable areas. Instead of taking into account concerns voiced by residents and allocating resources to resolve these issues, problems in poor areas
(outside of the city's business-oriented target areas) were overwhelmingly
ignored.

Coincidentally, the crack epidemic diverted attention from long-standing
issues with housing and neighborhood conditions by redirecting the public to
concerns about the morals and habits of poor people. Accordingly, declines in
demand for labor as companies fled to cheaper locales with far weaker environmental regulations were used to justify pro-business environmental policies
that have, ultimately, done nothing but disproportionately harm working-class
communities and further perpetuate the health disparities of their residents.

Throughout the 1980s and early 1990s, the City of Flint did make a few attempts to improve deteriorating neighborhoods by engaging in programs
designed to revitalize its downtown and address environmental problems in

targeted areas. Unfortunately, even these minimal efforts to impose controls on toxic exposure failed to take hold. This had a lot to do with the fact that the city de-prioritized these issues, even as citizens were ramping up efforts to remove toxins from their neighborhoods. City officials, most notably Mayor James Rutherford in the early 1980s, deprioritized neighborhood revitalization efforts and established a pattern of being dismissive and neglectful in handling citizen complaints. Residents were forced to take their environmental concerns to the state—which had also developed a habit of sidelining them.

After a few decades of benign neglect, Flint's minimalist approach to providing essential services—as well as its approach to managing public discontent regarding environmental hazards—has become institutionalized and increasingly apathetic and detrimental to the population's health. There was a time when residents' quality of life seemed to matter more and the city actually made efforts to gain citizens' loyalty and trust by providing reliable public services and being responsive to their concerns.

But, as depopulation escalated, property tax revenues continued to plummet, and poverty became more of a norm throughout the city, the decline in the quality of public services escalated and the pleas from citizens urging authorities to address these service problems were for naught. This transition is best exemplified by the city's lax efforts to protect residents from overpriced and polluted drinking water, which will be the subject of the chapters that follow.

3

Turning a Cold Shoulder

The Government's Response to Citizen
Complaints about Flint's Water

Flint's most recent troubles with drinking water quality were triggered in 2013. Emergency manager Ed Kurtz, who had held the same role from 2002 to 2003, authorized a switch in the city's water source from the Detroit Water and Sewerage Department to the Flint River. The contract with the Detroit department was up, and reportedly the switch was a cost-saving measure. The Michigan Department of Environmental Quality had recently assessed Flint River water and found it highly susceptible to contamination. But Flint's state-appointed officials claimed the shift in water source was in Flint's best interest.[1]

With glasses of treated water from the Flint River raised in the air, about a dozen local officials, including City of Flint council members, Mayor Dayne Walling, emergency manager Darnell Earley, and representatives from the DEQ, gathered on Friday, April 25, 2014, to celebrate the city's liberation from Detroit's water system.

Howard Croft, the middle-aged Black director of Flint Public Works, welcomed the audience of public officials and city employees to the ceremony for this momentous occasion. He spoke with great anticipation about the city's plans to stop the intake of water from Detroit. Flint was ending a fifty-year relationship with Detroit and had plans to temporarily use the Flint River until the city's primary water source, the Karegnondi Water Authority pipeline, was built. That pipeline would draw untreated water from Lake Huron to Genesee County.

After passing the final bacteria test for Flint River drinking water the previous Friday, Flint was permitted to close its connection to Detroit's thirty-six-inch water main. To reassure residents concerned about the change in water sources, Darnell Earley said, "When the treated river water starts being pumped into the system, we move from plan to reality. The water quality [will] speak for itself." Water treatment officials explained that extensive updates to Flint's water treatment plant over the previous nine months gave them comfort that the Flint River was "a different river than it was the last time [they] used it"—which had been before the Clean Water Act.[2]

With Mayor Walling's push of a small black button, Flint's water started coming from the new source. "Individuals shouldn't notice any difference," said Steve Busch, district supervisor in the DEQ's Office of Drinking Water and Municipal Assistance, to *Flint Journal* reporter Dominic Adams. Officials predicted that it would take two days before Flint residents would notice any change in quality in their water.[3] Despite the DEQ's assurances, though, complaints from residents began to pour in documenting concerns about the foul-smelling, discolored water flowing from their faucets.

Via Freedom of Information Act requests to the DEQ and the US Environmental Protection Agency, I obtained emails from Flint residents addressed to President Obama and the White House dated 2014 to early 2016 that offer vivid descriptions of the problems that residents were facing. For instance, one Flint resident complained on October 2, 2014, "Since switching to Flint River, my water stinks like sewage at times. . . . Brown sediment in sink basins and kitchen sink stays backed up." People in Flint began reporting hair loss and rashes to local water regulators.[4] Residents were quick to communicate their concerns, yet little was done to address or even acknowledge the fact that the plan to use the Flint River as a drinking source was obviously deeply flawed.

In this chapter, I document Flint residents' fears and complaints regarding the State of Michigan's decision to change the city's public water source from treated water purchased from the Detroit water system to the Flint River. I also chronicle the efforts and institutional circumstances that enabled local, state, and federal regulators to downplay the drinking water problems and undermine opportunities to resolve residents' very real concerns.

Residents' Problems with the Flint River

The Flint River's History of Pollution

Flint residents' primary opposition to using the Flint River as their water source had a lot to do with the river's long history of pollution. Since the late 1950s, the river had been dirtied and made dangerous by extensive industrial pollution, often via contaminated tributary streams.[5] Even then, city officials and locals knew that companies like General Motors and Buick were sending industrial deposits into the sewage system.

Buick, for instance, dumped oil, acids, chromium, and other waste materials into the river over the course of several decades. In 1967, as oil began to pour from the drains and into the river near Buick plants (especially the foundry), the City of Flint obtained a survey of storm sewers conducted by the city's Water and Air Pollution Committee. After reviewing the survey, the city demanded that Buick stop polluting the Flint River. Buick cleaned up the mess, but this incident wasn't the river's first or last problem with pollution.[6]

By the early 1970s, discharges from the AC Spark Plug Division were significantly affecting water quality in Gilkey Creek, which flows through residential and park areas in Flint. The creek carried a metallic by-product containing nickel and trivalent chromium that significantly influenced the alkalinity of its water. At times, people reported seeing a bright orange-red discharge (which contained high levels of sulfates, iron, and zinc) in the creek. Other pollutants came from garages and gas stations that routinely dumped oil, gasoline, and steam-cleaning waste in Gilkey Creek. Again, the river's alkalinity peaked and bacteria levels were high.[7]

Decades earlier, the State of Michigan's Water Resources Commission had ordered the City of Flint to address the problems with Gilkey Creek. In 1955, for example, the commission required the city to enforce an agreement with AC Spark Plug regarding the water. According to commission field agent Robert Parker, AC agreed to stop dumping anything but cooling water into the creek. However, 136 nearby homeowners and the city and county's health director, Dr. L. V. Burkett, signed a petition claiming that the company was still depositing oil, cyanide, and other toxic wastes in the creek. The city was involved because these waste streams were not confined to Gilkey Creek but passed through a city sewer.[8]

In early 1960, state water officials in the Water Resources Commission called for the immediate reduction of pollution in the Flint River. They also began to push for the state to find another water source for Flint residents due to the city's inability to control pollution within its current water source. The city had eight industrial and three municipal sewage treatment plants, but it was failing to control the pollution in the Flint River.

A 1959 study by the Water Resources Commission and the Michigan Department of Health found that the river was "grossly polluted for a fifteen-mile reach below the City of Flint." This survey also illustrated that the amount of pollution was critical downstream, near the treatment plant, and concluded that "it will be necessary to abandon the use of the river for municipal water supply to allow this entire resource to be utilized for dilution water alone." Officials warned that unchecked river pollution not only would affect Flint but also would pollute the Saginaw River and impact neighboring communities like Flushing, which had recently built a sewage treatment plant in an effort to control its contribution to river pollution.[9]

City of Flint officials ignored these warnings. On April 28, 1960, the Water Resources Commission summoned officials to Lansing to discuss plans for expanding the sewage treatment plant. The City of Flint was cited for its failure to control the Flint River's pollution, and a deadline was set to fix the problem: river pollution had to be significantly lowered by 1963. Specifically, the Water Resources Commission ordered the city to rebuild its overworked and underperforming sewage treatment plant and make progress toward reducing pollution in the river, "whereby the content of bacteria, human and industrial wastes will be restricted to the extent necessary to prevent conditions in the river injurious to public health, public nuisance, and aquatic life." The City of Flint responded by hiring the Boston hydraulic engineering company Metcalf and Eddy as a consultant to develop plans for improving the disposal system.[10]

The pace of the city's efforts, however, dampened any initial enthusiasm. A lawsuit filed by the Michigan Water Resources Commission on April 21, 1961, asked the circuit court to compel Flint to satisfy the mandates outlined in the commission's July 28, 1960, order. The commission contended that the City of Flint knowingly dumped human and industrial waste in the Flint River to a degree that was "so offensive to sight and smell as to create a public nuisance." City attorney William J. Kane, who filed the City of Flint's answer to the commission's lawsuit, acknowledged that there "may be some pollution of the river," but he rejected the claim that the pollution required the state to get involved and the city to comply with the commission's order.[11]

Although the Water Resources Commission had statutory authority to control water pollution in Flint and other Michigan cities (Act 245, PA 1929 as amended by Act 117, PA 1949), the City of Flint had to be compelled to finalize plans and complete the construction of the new sewage treatment plant. Local government in Flint was slow to comply with the commission's order despite the clear effects of water pollution that emerged: an unusual spike in the number of hepatitis cases in Flint and Genesee County in 1961 was traced to polluted county sewage drains and raw sewage in the Flint River, and dead fish were spotted floating in the Flint River after a 1962 chemical spill.[12]

Part of Flint's problem with the order to fix the sewage treatment system was shaped by the city's desire to find alternative solutions for its concerns with the water source. For instance, during this time the City of Flint began examining the possibility of using a pipeline to Lake Huron to satisfy future water needs. Despite pushback and delays, the city ended up spending nearly $8.2 million to increase capacity and modernize the sewage treatment plant that had opened in January 1964. Nevertheless, Flint began purchasing treated water from the Detroit Water and Sewerage Department in 1967.[13] It looked like the city would stop trying to clean up the Flint River and simply switch water sources.

But in 1968 the Water Resources Commission made additional recommendations concerning the pollution in the Flint River, and in 1970 the commission implemented a policy to raise the water quality standards high enough to allow boating and fishing. Yet even as regulations set by the Clean Water Act of 1972 and community groups like the Flint Environmental Action Team launched a "people's campaign" to clean up the waterway, the City of Flint continued to struggle to control pollution in the Flint River.[14]

By the late 1970s there was ample reason to be wary of Flint River water. Along with all the incidents outlined above, pollution was being found floating in or near the river and sewage system. Businesses were receiving astoundingly low fines for dumping large quantities of gasoline and other toxins into sewage systems and the Flint River. One of those companies, Thrall-Major Oil, pleaded not guilty in January 1979 to charges filed by the Genesee County Prosecutor's Office over alleged violations of state environmental laws. Thrall-Major faced fines of up to $25,000 for dumping 1,000–2,000 gallons of oil into Gilkey Creek and the Flint River in November 1978 and for dumping a total of 12,000 parts per million of oil and untreated industrial waste into a sewer manhole near St. John Industrial Park in December 1978. (The legal limit of oil was 100 parts per million.) Its pollution was deliberate and ongoing, yet Thrall-Major was eventually fined just $1,000.[15]

Days after Thrall-Major's plea was entered in the Flint District Court and Judge William S. Price III imposed the paltry fine, Flint Township's fire officials began receiving complaints from nearly twenty residents on Mission, Burnell, and Walton Streets who could smell gasoline in their homes. The source was traced to a leak of nearly 1,900 gallons of gasoline from a malfunctioning pump at the MSI gas station on Flushing Road. The fire department confirmed that the pump had been leaking into the sewers for at least twenty-four hours before complaints from residents emerged.[16]

By the 1990s concerns about lead in the Flint River were mounting. A 1991 report by the Michigan Department of Natural Resources revealed that the river's lead levels exceeded state water quality standards. Silver and mercury were overly high, too. In the late 1980s it was discovered that mercury emissions from the Flint wastewater treatment plant were poisoning the river's plants, fish, and other aquatic life. While improvements were made with local industries using their own wastewater treatment plants for industrial pollutants and with the discoveries and subsequent cleanup of oil spills and leaks from underground gasoline tanks throughout the 1980s and 1990s, the reality is that heavy metals can remain trapped for decades in riverbeds, polluting the water that passes over these contaminated sediments.[17]

The City's History of Deception

Flint residents had other reasons to be critical about the change in their water source: city officials had long made false claims about the quality of their public drinking water and failed to inform residents about potential health hazards.

For instance, on October 10, 1977, the City of Flint temporarily switched from treated water to Flint River water when main lines needed repair. But officials never informed residents. The *Flint Journal* broke the news by publishing locals' complaints about foul-smelling water and stomach problems after October 12. The city switched back to treated water on October 14, although it would take almost ten days for the river water to be flushed out of the pipes.[18]

As in other poor and mixed-zoned residential areas, Flint residents also had to contend with the state's and the city's persistent denials about the health effects of pollution. News stories and reports on water pollution illustrated a clear pattern of state- and city-level bureaucrats acknowledging that drinking water was polluted, then immediately proclaiming the polluted water safe for human consumption.

One notable example came in May 1986, when chlorine-immune bacteria were found in Flint's drinking water. Fred M. Germaine, Genesee County's director of environmental health services at that time, announced in the *Flint Journal* that "without qualification, the Flint water supply is safe to drink." Then, in April, Dr. George P. Grillo, the county's health department representative, advised local water treatment professionals to increase chlorination from 3 to 4 parts per million (three times the average level of drinking water chlorination).[19]

Soon, the coliform bacteria were undetectable in water coming from the Port Huron plant, which the City of Detroit operated, but reports of coliform in animal waste in the water on Ballenger and Flushing Roads raised red flags. The state didn't know how the coliform had gotten there. Tests of Flint municipal water in multiple sampling locations continued to indicate low levels of coliform bacteria evidencing disease-causing organisms throughout June.

William Ewing, Flint water superintendent, described these bacteria as among "the strongest organisms in nature" and acknowledged that it was understandable that citizens were fearful of the water. Still, he consistently claimed that the water was safe to consume.[20] Meanwhile, hundreds of locals started purchasing bottled water. At first, the people buying bottled water were just those who disliked the taste of the increasingly chlorinated water, but more people followed suit when the media began reporting the city's trouble managing bacteria. Michael Rieman of Crystal Water Company in Burton said of the increased business, "They drove us crazy last week," and "We had to work late a couple of nights just to keep up with the demand."[21]

Mayor James Sharp announced on June 20, 1986, that Flint would conduct a system flush to clear out the bacteria. Officials hoped the process would be complete by mid-July. The source of contamination in water purchased from the City of Detroit was not identified, but experts believed bacteria were stuck somewhere in the distribution system. They were confident that flushing the system with sufficient pressure would solve the problem.

The city planned to start the flushing in the northeastern part of Flint, then move to the northwest, southeast, and southwest quadrants. Residents were to be informed before the process started in their neighborhoods. Once flushing began, people would need to avoid washing white clothes because the flushing would result in a higher amount of rust in the system, discoloring the water. Again, the city assured its residents that the water was safe to drink and that, according to the Genesee County Health Department, no "discernable increase" in illness could be attributed to the water.[22]

Citizens eventually learned from state records that the city had been battling the pesky bacteria for two years, not two months. Water officials, including James K. Cleland, deputy chief of the public water-supply division of the Department of Public Health, were operating under the assumption that "every system seems to have some level of bacteria." When Flint's water tested positive for bacteria in January, September, November, and December 1984 and again in March, June, and September 1985, they kept it quiet.[23]

According to Superintendent Ewing, the city hadn't shared those problems with the public and applied no special treatments because "it was not the problem it was this time." Officials insisted that earlier contamination had been discovered and corrected within a few days. The current problem, however, involved bacterial contamination more than ten to fifteen times the normal levels, and it was proving difficult to control. Ewing recalled that no public notice was required when the city detected more substantial amounts of bacteria in June 1981 (and temporarily treated the water source with chlorine). After the water distribution system was flushed, discoveries of bacteria actually did decrease substantially.[24]

Flint began reducing the water's chlorine level the first week in August by .5 part per million (down to 2.5–3 parts per million, which is two to three times higher than the average amount of chlorine added to municipal water to kill bacteria). Three weeks later the bacteria reemerged. Rather than implement "drastic" measures like calling on residents to stop consuming the water or boil it before use, the city continued its aggressive chlorine treatment regimen and looked for—but never found—the source of the bacteria. Like Grand Rapids, Muncie, and Springfield, Flint was not able to pinpoint the source of the bacteria contamination.[25]

The State of Michigan, under Governor James Blanchard, wasn't helping. By October 1986, Mayor Sharp had requested but not received state help locating the source of the bacteria, even though the contamination clearly was not limited to Flint or even to Genesee County. Water from Detroit's Port Huron treatment plant was contaminated, and bacteria had been discovered in Macomb County. In his plea to Governor Blanchard, Sharp noted that "the problem has gone on far too long . . . and our citizens have been subjected to drinking a water supply that is distasteful." Still, without any direct evidence that Flint's contamination was originating anywhere other than in the city, the governor's office did not approve the requested manpower. According to William Kelley of the State Department, "I don't think there's anything new or different we can do" to find the contamination source and protect the nearly 200,000 people in Flint and Genesee County from illnesses caused by the presence and treatment of bacteria in the drinking water. The bacteria continued to be controlled by fluctuating and high levels of chlorine. Again, the source of the pollution was never discovered.[26]

State-level water interventions have ebbed and flowed over time. Governor Blanchard's comprehensive strategies in the 1980s included attempts to hold industry at least somewhat accountable for its pollution. Under Governor Blanchard, state laws were passed to ban the use of pesticides. It is also the case that, on September 18, 1984, Governor Blanchard unveiled his plan to protect Michigan's groundwater from pollution. His Council on Environmental Protection spent over eight months developing the plan, which responded to the problem of leaking underground storage tanks and landfill pollution in groundwater. The council recommended a number of amendments and rules that would help the state develop standards for storing hazardous materials, consolidate state laboratory water testing procedures, stop landfills from being located near usable aquifers, create a state fund to pay for alternatives to landfills, and compel any person or business polluting groundwater to provide emergency drinking water to residents affected by the pollution.

Blanchard's environmental advisor described this initiative as "a strong priority" on the governor's agenda, and the plan represented the state's willingness to take seriously its role in protecting the groundwater and the health of the 9 million Michigan residents who depended on it.[27] However, except for advocating for full cleanups of contaminated sites,[28] the vast majority of Blanchard's plans could not come to fruition before his 1990 ouster by newcomer John Engler.[29]

Under Governor Engler, the environmental community began to see a steady decline in communications and relations with the state. Engler characterized Michigan environmentalists and organizations like the Department of Natural Resources as top-heavy and overly responsive to the "environmental cris[es] of the month." His administration's restructuring of the state's response to pollution involved attempts to streamline the DNR by eliminating departments like the Well Drillers Advisory Committee and installing the Michigan Environmental Science Board, which personally advised the governor on a host of pollution issues from mercury contamination to discharge permit violations.[30]

Just as the state began to lose momentum on environmental protections in the late 1980s and early 1990s, environmental groups in Flint suffered the same problem. Many were struggling for members amid economic decline. Broke and perpetually unemployed, Flint residents who had once been active participants in environmental causes had withdrawn from the movement. The Flint Environmental Action Team, the city's oldest environmental group, announced its dissolution in 1991. It buckled under accumulating debts and diminished public interest in recycling. Other environmental groups, including SAVEIT, were reporting burnout, with members unable to keep up with the work and dedication required to participate in and lead interested individuals. Leaders could easily spend thirty to forty hours a week advocating for a program or legislation. These labors were often volunteer work, and in the late 1980s, it was becoming apparent that unpaid and overworked volunteers were no match for corporations willing to combat every fact and figure with data presented by well-spoken, company-financed researchers and lobbyists.[31]

Recent Issues with the City of Flint's Drinking Water

Affordability

A few years before anyone outside Flint was aware of the city's water equity issues, Flint residents participated in marches and regularly showed up to city council meetings to register their concerns about the affordability of their public drinking water.

Although Flint council rules adopted during state receivership required residents to limit public comments to five minutes, residents used their allotted time compellingly and consistently to express outrage at the cost of essential water and sewage services. As one citizen noted at a July 28, 2014, council meeting, "The water bills are atrocious. . . . People are paying $140 a month when once upon a time they were only paying $60 a month. Water is a necessity. The human body is 75% water. You can go without food, but you can't go without water. And I tell you this much, if an elder died in their house for lack of water, which is dehydration, to me that's a crime. That's murder because you cannot strip somebody away from the natural necessities of life."[32]

Flint residents had good reason to complain about the cost of public water. Between September 16, 2011, and July 1, 2012, the billings for all retail customers of Flint's water and sewer department increased by 35 percent, a hike suggested by the DEQ to subsidize efforts to repair the city's declining water distribution system. On December 1, 2011, emergency manager Michael Brown, who was appointed by Governor Rick Snyder, implemented an additional 12 percent increase in water rates effective July 1, 2012.

In attempts to advocate for clean and affordable drinking water—and to protest the Michigan State Police Force's excesses in helping the Flint Police Department patrol streets and investigate crimes—Flint residents, supporters, and city council members took to the streets outside city hall on Monday, July 14, 2014. While city leaders, including Councilman Eric Mays, took turns making remarks with a bullhorn, a small group of residents and supporters, both young and old, stood on the sidewalk and lawn in front of city hall holding signs that read, "No More High Water Bills" and "E[mergency] M[anagers] Go Away." Former Flint councilman Wantwaz Davis, who organized the march, said the citizens deserved an end to the exploitation. "We're really concerned about the mistreatment of the community," Davis said. "We're teaching these young people to respect the system, but the system is failed."[33]

Unfortunately, protests throughout the cold winter months of 2015 and the multiple meetings where residents shared their tears and frustrations with local leaders did not change the cost of water in Flint. By early 2015, some residents' water bills were over $300 per month, and frustration was coming to a head.

During this time, there were also plenty of residents speaking about how unfair it was that their water rates were higher than those in neighboring communities. These circumstances led some residents to form organizations like the Coalition for Clean Water, which formed in a church basement in the spring of 2015. Meanwhile, other residents focused on specifying how water costs compared between Flint and other cities or municipalities in Genesee County. Jerry Preston, a white male and former president of the Flint Area Convention and Visitors Bureau, showed up to a January 13, 2014, city council meeting to share his research findings. "What I've left on your desk today is the comparison of two water bills, Flint vs. Flint Township, and it's kind of interesting." He went on to say,

It's my [City of Flint] water bill vs. the one I've gotten in Flint Township. The amount of the water [bill in Flint] is 15 times the Flint Township bill, and the Flint bill is just one unit. . . I went back and recalculated these bills using the Flint bill on the Township rate, and the Township bill on the Flint rate. And you can see that no matter how you look at this thing, our cost of water is 2 to 3 times the cost of water in Flint Township. Now, when you recognize that the city of Flint sells water to the County, and the County sells the water to the Township, and there's billing and other markups all through there, you kind of wonder, "How's that possible?"

Due to the high costs of drinking water, many Flint residents were unable to pay their skyrocketing bills. Their water accounts went delinquent, which in turn rendered them unable to secure housing. Property owners soon found themselves having trouble getting renters; people couldn't open water and sewage accounts because they were delinquent on old bills. A landlord at the same council meeting commented,

Water deposit for tenants and the water rates are driving the people away from the city because the tenants can't get the water turned on. I keep a log of the people that call me when I have a house for rent. Properties usually rent in about a month, after about 20 people have called and maybe I've shown it to 5 or so people. Now it's more than triple that number. It used to be a month or two, maybe, it would be vacant. Now it's 4 to 6 months and still going. I've done what I can; I've reduced the rents, I've even taken only the security deposit to move someone in, with the addition of a receipt for them getting the water turned on in their name.[34]

The average Flint water customer was paying $150 per month—among the highest rates in the country. In April 2015, 378 customers received shutoff notices due to unpaid bills.[35] The city sent 5,503 lien letters totaling $3,396,712 to property owners with water and sewer charges that were over six months past due.[36]

Flint residents understandably saw themselves as everyday people getting a raw deal. That sense deepened, especially after lien notices were sent for delinquent water bills throughout 2017. Residents suggested that in many instances, broken meters had affected their charges. In turn, as one resident noted at a city council meeting on May 17, 2017, "It's entirely understandable that residents would feel indignant about being asked to pay for such water. . . . I'm sympathetic to the city's financial predicament. I know that often requires making tough choices, but in deciding how we're going to fund the city's services, we also have to decide what kind of a city we want to live in. Is it a city whose revenue stream is maintained through threats and coercion?"

Water Quality

Not only were residents grappling with water affordability and tax liens due to delinquent charges, but they were also still deeply concerned about the water quality.[37] Following the shift away from the Detroit Water and Sewerage Department in April 2014, Flint residents began issuing complaints that echoed those lodged when the water had been shifted briefly in October 1977: the water was discolored and foul-smelling; bathing in or consuming the water caused hair loss, stomach problems, and skin rashes; and families were forced to use bottled water for everyday tasks such as cooking, bathing, and washing their hair. Flint residents argued that they should not have to purchase water unfit for use.

At a November 19, 2015, city council meeting, Bethany Hazard, who lived on the west side of Flint and had been a city resident for just over a decade, reported that she had noticed quality problems with the water as soon as the water source switch happened. She said she did not understand why residents were being required to pay for dirty water and had to endure minimal response to their concerns.[38] As she put it,

> I had yellow sediment in my water for months over the summer. I called and went to more than one office for help. It took months for the city to come out to my house. . . . I was taking a bath with yellow, sandy sediment with black flecks in it. . . . This is outrageous. I've seen friends move out of my neighborhood because of the price of the water and also because they don't send the police to my neighborhood for some odd reason. So, I hope that somebody will listen. If we're paying these high service fees somebody should be out there right away to see what's going on with the water.[39]

Since the local water regulators were ignoring expressed concerns, low-income families and the elderly primarily counted on each other and on community groups that stepped up to help. Schools were asking parents and community partners to donate bottled water. A Flint resident wrote to President Obama on November 23, 2014, that citizens had to work together because families were struggling and "the city [could] not provide affordable, safe [and] usable water to citizens." The letter writer was a volunteer at an emergency water site and had observed that families who could not afford the water were attempting to make good use of the resources available to them, including "using buckets carried from an open hydrant for sanitation." At the time, the emergency water site was equipped only to pass out water. It could not provide shower or laundry services because the "water is even off there [at emergency water site] due to egregious expense." Ultimately, this letter writer called for public action in Flint:

> The water problem in Flint should be addressed as a public health emergency because families without water are considered homeless. Households without water access are in danger of loss or damage to the home and having children separated from their parents by the social service system that is suppose[d] to help them. No one can stay healthy or be a productive citizen in this society when they have to spend time looking for access to something as fundamental to survival as clean water to drink, running water to wash the body in preparation for worship, work or school.[40]

Emails addressed to President Obama commented on the lack of response from local officials. Flint residents were especially upset that representatives had failed to notify them *before* issues with the water started. Instead, one resident wrote, "they raised [our] rates and waited months. As this proceeded, [Flint residents received] several different boil advisories due to fecal bacteria in our water. We, the people, are fed up and disappointed in our state and city leaders." In a handwritten letter to the president, another resident noted that "the local politicians are all about saving money. Our safety and health should be the focal point. While they all jockey for position, we are still drinking unsafe water."[41]

Many people felt that poverty was the biggest reason local and state-level officials were uninterested in addressing Flint's concerns. As one resident recalled, "My life was going very [well] until G.M. moved their factories out of Flint, MI. The city became desolate. We have low paying jobs, high utilities, high rent, and no chance of a fast recovery." Another, emailing the White House on January 27, 2015, painted a picture by saying, "I live in a city where crime is high, schools are closing, neighborhood[s] are declining and [there is] very little law enforcement. It seems like officials and the people who run our city do not care about the people who are trying to be upstanding citizens and make it in an already impoverished environment."[42]

A consistent theme in the complaints I secured from the DEQ and the EPA was that Flint residents could not understand why they had to pay a high price for water they couldn't consume or even bathe in. Before free bottled water and filters were made available to Flint residents for a limited time, one resident said, "We have to purchase it on a daily basis, and it's not in our budget. We don't have enough income to buy water like this."[43]

State regulators also heard citizen complaints. Richard Benzie, the DEQ's field chief of operations, informed the EPA Region 5's Thomas Poy, Ground Water and Drinking Water Branch chief, and Jennifer Crooks, Michigan program manager, that an "unidentified individual" had called to speak with the person responsible for ensuring the safety of Flint drinking water. Benzie mentioned that he had provided the names of the district supervisor and district engineer who oversaw the water distribution system, but the caller refused to accept the information because "he wasn't going to waste his time with them." Benzie was giving the EPA Region 5 supervisors a heads-up: someone who refused to identify himself and follow the chain of command would likely be in touch, wanting to speak with someone in Washington, DC. Wishing them good luck with the "adamant" caller who voiced distrust for state and regional regulators, including Jennifer Crooks, Benzie personally suspected Crooks had already spoken to the caller.[44]

FOIA records demonstrate that Flint residents were spot-on in their distrust of water regulators. Some water treatment administrators had even purposefully misled Flint residents who contacted them for guidance and resolution during the water crisis. Jennifer Crooks was one prime example.

On September 14, 2014, Robert Bincsik, Flint's water distribution and sewer maintenance supervisor, forwarded to EPA Region 5 a worried Flint resident's complaint to the City of Flint Water Department about a potential Safe Drinking Water Act violation.[45] According to the resident, there was a problem with Flint's water that local officials were ignoring despite persistent complaints. "My Flint City tap water is brown. It cannot possibly be safe to drink," wrote the resident. The complaint went on:

> No person is available to discuss this issue with me, even after already having three boil water advisories in less than two months['] time. There has been no announcement made of this current issue, nor has there been any advisory to not drink the water. This has not been addressed or even acknowledged by the city. My neighbor has been having the same issue, and she has also not received any answers, acknowledgment, or regard for her lack of safe drinking water. It is disgusting, and they are not even informing the citizens that it is happening.[46]

Upon receiving the referral on September 16, Crooks was asked to "provide a response" indicating that the issue was a "valve issue and that the water flow should improve."[47] Crooks contacted the DEQ's Stephen Busch, district supervisor, and Mike Prysby, district engineer, just before noon on September 17 for guidance. "Yep," she said, "[we] have another Flint complaint. Her water at the tap is brown as of this past Sunday. I thought Flint was flushing in the areas where the valves were closed? Not sure if this woman's address is close to that area? Have there been other complaints of brown water?"[48]

After conferring with the City of Flint's water treatment plant and the Water Service Center regarding the "latest complaint,"[49] Prysby informed Crooks that the resident did not live in the boil water advisory area. He thought they should "hold-off on calling her" until they were clear about Flint's water treatment plant's response.[50] Bincsik told Prysby via email that the query was from a home south of the boil advisory area and that the water issues there were likely the result of improper hydrant use by nearby contractors. "I will have someone go out and flush the hydrants near this address," Bincsik wrote to Prysby and Brent Wright, water plant superintendent.[51]

Crooks took Prysby's advice to hold off on responding to the complaint. A new complaint came in later that day from the same area: cloudy water.[52] Just after six in the evening on September 17, Prysby wrote Crooks again. He wanted her direct phone number: "I wish to call you."[53] A little more than an hour after they'd spoken on the phone, Crooks contacted the first resident to assure her that water problems were likely due to a malfunctioning valve. She wrote that "most likely the cause of brown water [was] an incorrect use of the hydrants by demolition contractors in the area during the time. . . . The contractors don't always let the water flush for 10 minutes[;] so they turn off the hydrant, and the sediment is still in the main. Your distribution lines were flushed this afternoon, so your water should be more clear now." Crooks encouraged the resident to get in touch with her if she needed further assistance, adding, "I have several contacts at the MDEQ that are very actively assisting the City of Flint with their drinking water issues." The resident immediately responded, expressing immense gratitude: "I cannot tell you how much I appreciate knowing that someone is doing something—or at least advocating for answers for us if nothing else."[54]

But Crooks wasn't telling the whole story. EPA FOIA records demonstrate that she had learned more in her private call with the DEQ's Michael Prysby than she shared with the resident. In one email, Crooks updated her Region 5 colleague Thomas Poy about the Flint water complaint, indicating that water problems in Flint weren't "going away anytime soon."

> I talked with Mike Prysby at length about the future of these drinking water problems in Flint. He said this next problem that will rear its ugly head is the HI [high] TTHM [total trihalomethanes] problem that is now surfacing. Mike just sent off a compliance letter to Flint this week about the high TTHMs. Seems that the plant is able to get the TOC [Total

Organic Carbon] levels down to an acceptable level, but due to the fair amount of precursor/organic loading from the Flint River, including the farm runoff from the entire watershed, they're still seeing high TTHM levels. Of course, the corrective action is to connect with Lake Huron water in 2016—which is 16 months away. But in the short term, Mike said, "Bottom line, can these people drink the water now?"[55]

As federal water regulation officials struggled with how best to follow up with Flint water concerns after the switch to the Flint River source, the DEQ was trying to get Flint to deal with the issue of high TTHM in the water.

The EPA's Stage 2 Disinfectants and Disinfection By-Products Rule of 2006 requires public water suppliers to test drinking water systems for the disinfectant by-products TTHM and HAAS (halo-acetic acids) on a quarterly basis. The City of Flint was measuring by-product levels from eight sampling sites. Testing results indicated that six out of eight sampling locations exceeded the maximum contaminant level for TTHM in the first quarter of 2014 (collected on May 21, 2014).

By the second quarter (August 21, 2014), each site tested far surpassed the maximum contaminant level for total TTHM. Lockwood, Andrews and Newman, contractors hired to assess the issue, blamed the TTHM problem in Flint on a confluence of factors: sewer leaks that increased total coliform levels, a malfunctioning ozone system that increased chlorine feed, unlined iron pipe, and broken valves resulting in stagnant—and thus bacteria-prone—water throughout the city.[56]

First Wave of Crisis Response: Culture of Cover-up

Flint's sluggish response to water quality concerns persisted until January 2015, when the city was legally required to notify the public about its first major water emergency involving toxic TTHM. In a notice titled "Important Information about Your Drinking Water," Flint announced to residents that TTHM detected in city water exceeded the maximum contaminant level of 80 µg/l (micrograms per liter). "Our water system recently violated a drinking water standard. Although this incident was not an emergency, as our customers, you have a right to know what happened and what we are doing to correct this situation," the notice stated.[57] The city reported that ingesting drinking water containing excessive levels of TTHM did not require boiling. So citizens, especially those without a "severely compromised immune system," believed that water right out of the tap was safe. To apprise people on its progress managing the TTHM issue, the city posted updates on its website and scheduled public meetings, again allowing residents to believe the city was properly managing the water troubles.[58]

Flint distributed its first notice to residents regarding TTHM on January 2, 2015, yet EPA FOIA documents illustrate that, by the end of March 2015, water treatment regulators were still wrestling with which populations their public notices should identify as vulnerable to TTHM.[59] Jennifer Crooks had realized that pregnant women were excluded from the list of vulnerable populations and remembered the comment by Miguel Del Toral, EPA Region 5's Ground Water and Drinking Water Branch regulations manager, that TTHM could potentially cause miscarriages. So she asked the DEQ's Mike Prysby whether pregnant women could be added to Flint's public notice "so women [would] know there could be a risk and can make their own decisions."[60] A few days later, Crooks wrote her EPA colleague Poy about the suggestion, but the DEQ would not require the City of Flint to add pregnant women to the notice because "the actual mandatory health effects language does NOT include 'pregnant' so they didn't feel they could require the City to add 'pregnant.'"[61]

While Crooks reminded Prysby that pregnant women were included in the public notice template for TTHM provided to the state, Prysby retorted that the state was not mandated to implement suggestions from the Public Notice Guidance Manual.[62] Although public notices are required to include populations and subpopulations that are at risk if exposed to the contaminant, the EPA believed the issue was moot, since the agency does not have the power to compel states to comply with suggested protocols and public notices.

After water samples were collected on February 17, 2015, a Total Trihalomethanes Fact Sheet—which was created by the Genesee County Health Department and distributed with the water quality updates—informed pregnant women of possible risk. However, the state continued to refrain from identifying any specific populations at risk and opted in its public notices to describe in only the broadest terms the health effects of TTHM.[63]

Rita Bair, the EPA's Ground Water and Drinking Water Branch chief, asked for additional comments concerning her summary of the Flint discussion among water treatment managers and professionals at the EPA. In response, Crooks suggested on July 9, 2015, working with the state to fix the problem, because "rubbing their noses in the fact [that] we're right and they're wrong" would not best serve the Flint community. She wrote to the water treatment folks who attended the meeting,

> MDEQ did not tell Flint to maintain corrosion control; instead, they treated the change in source water as a new system. I'll bet that the State will take this personally since they are responsible for the City of Flint's

actions, which isn't a bad thing but they may get VERY defensive. We can get into the weeds. We want to work with the State, and the City to address ALL of the contaminant issues going on in Flint; bacteria, TTHMs, Legionella, lead—from a holistic approach to getting some form of corrosion control working that doesn't minimize treatment for bacteria or TTHMs. It doesn't make sense to discuss with the State what happened in the past; we need to move forward and work with the State as our partner.[64]

As I looked at the correspondence between EPA and DEQ water management regulators, it was apparent to me that state and regional water regulators avoided confrontation or conflict with the state at every turn. Though the EPA considered ways of partnering with the State of Michigan to resolve water quality concerns, FOIA documents from the EPA illustrate that the agency had long looked the other way when it came to the state's violation of environmental laws.

Every quarter, EPA Region 5 generates Enforcement Targeting Tool Scores and guides the Michigan DEQ in its enforcement of environmental laws. These data show that the US EPA consistently raised concerns regarding the state DEQ's compliance with 1991's Lead and Copper Rule, which makes municipalities responsible for testing areas for lead exposure, fixing problems, and notifying residents about any potential dangers. These concerns were always acknowledged but never resolved or adequately addressed. For instance, the US EPA instructed the DEQ to "incorporate rule revisions into state oversight and enforcement operations" and "maintain a database management system that accurately tracks lead and copper action level exceedances (sample data), violations, and milestone data." It mandated that the DEQ discourage noncompliance or incentivize compliance by placing on "6-month monitoring" any water system that was "repeatedly not conducting its annual or triennial lead/copper monitoring."[65] And it suggested that the DEQ hold water systems accountable by sharing with the EPA sample data for all medium and large water systems, as well as results from a small system that was violating Lead and Copper Rule regulations.

Furthermore, the state had to inform water systems about their concomitant obligation to conduct Lead and Copper Rule compliance monitoring between July and September, when children are known to have elevated blood lead levels. In response, the DEQ acknowledged the need to improve how it managed water systems, yet it asserted that many of the requests, including violation reporting and the sharing of milestone data, could not be met due to state disinvestment or lack of funding.

As stated in these documents, the "MDEQ does not commit to issuing or reporting violations for PWSs [Public Water Systems] who monitored, reported late or failed to issue Tier 3 P[ublic] N[otices] (State Disinvestment. See Att. B.)"; furthermore, "MDEQ does not commit to 100 percent reporting of LCRMR [Lead and Copper Rule Minor Revision] milestone data during FY 2014, but will continue to improve the data as resources allow."[66]

In 2014, the DEQ was still using disinvestment as an excuse for not fulfilling various monitoring and regulatory duties related to the Lead and Copper Rule (for example, Type 66 violations). In addition, the DEQ neglected to issue or report violations by water systems that reported late or failed to issue Tier 3 public notices about the presence of volatile organic and inorganic chemicals, including arsenic. The DEQ blamed its inability to adequately report noncommunity water systems' Lead and Copper Rule violations on a glitch in the web-based water management system, WaterTrack, which limited its capacity to input new violations.[67]

The DEQ's failure to follow up with known toxins in the water meant that many Michigan residents would consume water contaminated with arsenic, bacteria, and high levels of disinfectant by-products. Violations of the Safe Drinking Water Act in Michigan also illustrate that Flint was not the only municipality contending with these poisons. For instance, in 2013, fifty-four community water systems received seventy-five citations for failing to implement primary routine and follow-up monitoring procedures for total coliform. Twenty community water systems were cited for failing to implement routine lead and copper tap rules, and thirteen were cited for failing to provide Lead Consumer Notifications.[68] Astoundingly, thirty-three were cited for TTHM exposure.[69]

Of course, Michigan water systems are unable to eliminate *traces* of toxins, including nitrate, arsenic, and mercury, in the water. But in 2013, 482 out of 1,461—or about 33 percent—of the total number of regulated systems (community and noncommunity) received violations for nitrate; 64 were cited for significant monitoring or reporting regulations involving arsenic; and 29 were cited for monitoring or reporting regulations involving mercury.[70]

Consequences of Flint's Culture of Inaction

The City of Flint was learning it could not easily convince residents that the water problem had been resolved. Many rejected the city's stance that the TTHM violation did not constitute an emergency. They advocated for more transparency. As Flint officials were insisting that the only problem with the water was how it looked, residents and community groups—including Concerned Pastors for Social Action, the Democracy Defense League, Water You Fighting For, as well as other Flint groups and citizens—began staging protests. January, February, March, and June 2015 saw more and more protests. Activists held town hall meetings and press conferences and used social media and websites dedicated to the water crisis to attract international attention to Flint's problems.[71]

Flint mayor Dayne Walling, convinced by state regulators' insistence that the water was potable, tweeted on April 2, 2015, "[My] family and I drink and use the Flint water every day, at home, work, and schools." This was only days after residents had received notice that Flint remained in violation of the Safe Drinking Water Act. That is, despite residents' ongoing city, state, and federal complaints, officials in Flint *kept* saying the water was safe.

City administrator Natasha Henderson, authorized by emergency manager Darnell Earley to oversee the Flint Water Department, not only explicitly called water from the Flint River safe but also declared that the city was "doing everything possible to make the system more secure, efficient, and affordable."[72] However, two weeks earlier, on August 14, 2015, in an email to President Obama, one resident commented, "Our water in Flint, Michigan, is contaminated, and our bills have soared. Warnings were sent out, only about a year after the community had drunk and used it in other ways. My daughter's cat would not drink it, just to give you an example of the horror of it all. I am scared to wash my hair and take a bath. . . . I feel as if they are trying to kill us. Can you please help to solve this problem. Something is very, very wrong."[73]

People were connecting the crisis to especially devastating effects on economically fragile families and children. On September 15, 2015, a twenty-one-year-old mother of three wrote President Obama to plead for federal aid in resolving the water issue in Flint. She had not believed that the water was bad until the evidence of the issue—brown water—poured from her faucets. She feared for her young family, especially her two-year-old daughter, who had persistent health problems, and she wrote a series of questions to gain some insight on how best to keep her children "out of harm's way." "My middle daughter," she wrote, "has 9 specialists and currently no diagnosis. . . . She gets sick very easily. When a child is at risk for infection, you are advised to avoid anything that could cause him/her to get sick but when the risk is in their home water, how can you avoid it?"[74]

Their city was already vulnerable. Now, as these individuals' accounts show, it was being brought to its knees. Pleading with President Obama in May 2015, a local wrote, "It has been bad for a year . . . and it is not getting better. Please shine some light on this situation and help our city. Our city is so close to a breaking point that I fear this summer could get crazy. No water or jobs can make an already violent city get even worse."[75]

The Flint water crisis is not the first time residents there have faced water quality concerns. As demonstrated in this chapter, Flint residents have consistently contended with water contamination issues since the 1960s. In the early 1960s, Michigan's Water Resources Commission exercised its statutory authority in attempts to compel the City of Flint to control the pollution in the city's drinking water source, which was the Flint River. The City of Flint delayed efforts to improve the water because its interest in pursuing a pipeline to Lake Huron superseded its need to modernize its water treatment plant.

Over fifty years later, Flint's plans to build a pipeline to Lake Huron were finalized. But this time the State of Michigan, while governing in the place of Flint's elected officials, made this move on the city's behalf after the state's ongoing entangled relationship with the business community and perpetual disregard for public safety had neutralized the discretion of water regulators.

Subsequently, several developments occurred after the city switched its water source back to the Flint River, despite the river's history of industrial pollution and local dumping. Resident complaints about water quality peaked amid ongoing concerns about the costs of water. Residents repeatedly tried to force authorities to respond to their concerns about the cost and quality of their water. However, demands for clean and affordable water—which resulted in many protests and marches—did not solve the problem. The water continued to be expensive, and residents continued to worry that it was making them sick.

By persistently downplaying residents' concerns and disregarding evidence of existing problems within the water system, state-appointed leaders, as well as regional and federal regulators, betrayed residents by prolonging water issues and threatening to undermine coordinated efforts among Flint community groups. These officials, as we shall see in the chapters that follow, exploited conflicts among opposing allies and encouraged mistrust. Indeed, it seemed as though Flint was reaching its boiling point.

4

A Tale of Two Flints

Conflicts among Residents concerning the City's Water

The Flint "residents turned activists," as depicted by Benjamin J. Pauli in his book *Flint Fights Back* as "water warriors," engaged in action about their water not due to "prior political commitments but by the personal impact of contaminated water."[1] These residents who had witnessed clear changes in the taste, smell, and color of their water no longer believed that the state was doing its part to keep them safe. In response to this problem, they opted to use their collective power and agency to address concerns about water quality and affordability.

At the same time, other residents were not convinced that their water was impacted. They thought that fears about water contamination and concerns about "an obvious genocide"—spread by Black activist and then city councilman Wantwaz Davis[2] and others in Flint—were exaggerated. This was the case even after boil water advisories concerning coliform bacteria in late summer 2014 and the TTHM scare in January 2015 inspired the initial bottled water giveaways.

So, Flint's residents were divided. Residents wading through the obstacle course of misdirection and lies paved by state and local officials had developed factions and even opposing viewpoints when it came to the water—and, frankly, when it came to plenty else in the city. Rumors about poisonous water stood against officials' assurances that the water was safe. The people of Flint differed in their opinions about both the extent of the problem and, if it existed, who should shoulder blame.

On one side of the issue, especially before independent researchers had proved any problems with the water, many residents felt that medical doctors, water regulators, and state-appointed officials were involved in a conspiracy to poison them by contaminating the overpriced water. To address this issue, residents began sharing their concerns as early as May 2014 by participating in the Flint River Water Support Group as well as collectives like the Flint Water Class Action Group created by a Black mother in the community, Florlisa Flowers.

On the other side of the issue, some Flint residents were critical of allegations charging state and local officials with reckless endangerment. They saw no purpose in efforts to hold those officials accountable for issues the residents believed were related to the city's aging and deteriorating infrastructure.

Therefore, some residents felt betrayed. They were seen on television news on June 4, 2015, declaring their intent to sue and were quoted in print media as being suspicious of the state's intentions. But other Flint residents didn't participate in these efforts. According to Pauli, residents who were among "those whose water was clear, palatable, and easy on the skin and lungs were slower to conclude that it was a threat to them."[3]

I think of these opposing viewpoints when I recall interviews I conducted—each lasting over three hours—with Flint community leaders about the water crisis. These interviews, held in local Flint restaurants, occurred before evidence emerged verifying Flint residents' accounts about the water. The interviews, which included a range of residents, exposed distinctive but intersecting perspectives concerning the crisis. Since Flint still has active community groups and block clubs, I sought to learn about how these groups, both independently and collectively, were responding to the water crisis.

I contacted Flint Neighborhoods United, a collective of crime watch groups, block clubs, and neighborhood associations in the city. The collective takes a "city-wide perspective versus the focus of individual members of a specific neighborhood or area within the city" with the intent to share resources that promote change in Flint.[4] I reached out to this organization because of the work it does throughout the city to coordinate neighborhood advocacy and promote clear communication concerning key issues among different stakeholders in the community. In response to my outreach, I was put in touch with Bill Hammond.

A leader of Flint Neighborhoods United, Bill Hammond—white, retired, and middle-class—talked with me for hours about Flint, its history, its struggles, and its capacity to withstand obstacles. Hammond is a good-natured Flintstonian who bleeds community pride. As we munched on our food in a downtown diner near the University of Michigan–Flint, Hammond shared his motivations for doing community work. He had deep ties to the community, a knowledge of nonprofit and city-sponsored efforts to enhance civic engagement, and trust in what seemed to be reasonable responses from local and state officials to residents' concerns about the water.

Hammond's commitment to community-led collective action and his optimism concerning Flint's survival through tough financial times were inspired
by observations of his parents' devotion to interracial harmony and economic
justice back in the civil rights era of the 1960s. His parents, he told me, had
refused to be bullied and goaded into fearing Black people or moving to avoid
them. "My mother, in particular, really felt strongly about this," Hammond
shared, as he recalled the numerous rallies, protests, and race retreats his parents participated in during his youth. They advocated against the war in Viet-
nam and joined the fight against restrictive housing laws in Flint during the
Floyd McCree days.

The Hammond family was one of just two white families that chose to stay
on their block as it integrated. "We were the last white family that moved on
our street," he said. "That was when white flight all started, and so our neighborhood became virtually all Black within a matter of maybe three years. . . .
Within five to seven years, it was 90 percent [Black]."

His background led Hammond to a career in community development
and engagement. He retired from Salem Housing Community Development
Corporation due to illness and subsequently became committed to building neighborhood resilience and producing "our little paper," as he called the
neighborhood-based newspaper *Flint, Our Community, Our Voice*.

So, with a can-do attitude, Hammond believed that Flint's abandonment
was a reality that could be resolved by citizens' teamwork and dedication. As an
example, he talked about what residents had done when drug houses emerged
on a block: they took turns reporting the problem until the Flint police, who
had developed a reputation for showing up only to inquire about a dead body,
paid appropriate attention.

Recalling the families that had moved on and the still-vacant lots where
houses had been torn down, Hammond spoke in terms of being responsive to
the pain, loss, and grief in Flint. He noted, "For the neighborhoods dealing with
these huge losses they have to go through a process where they stop thinking
about their loss and instead start thinking about, well, this is where we're at
now. What can we do to keep going forward and make it better?"

Hammond spent a great deal of time discussing the virtues of neighborhood
plans developed by community residents and the Hub. The Hub is a neighborhood resource that helps community members address issues pertaining to
neighborhood upkeep by making tools like gardening implements available to
residents. Everything from rakes and shovels to rototillers and riding mowers
were available for residents to borrow. They just needed to show up—to the
Hub and for their neighborhoods.

I had contacted Hammond's organization, Flint Neighborhoods United, about the water. Just before arriving in Flint, I had seen a resident's video of a city council meeting on YouTube. In the meeting, citizens said that pets didn't trust the water and that it was killing people's gardens. When I mentioned this to Hammond, he noted that he had not personally observed issues with the water and thought the crisis was a play for attention. He referred to the water problems as "hype" that was distracting the city from the real issues: depopulation, poverty, and crime:

> There are people who just feel that it's just important for them to make as much noise about things as they can. The real story is there are in fact some real problems. . . . Detroit, of course, is experiencing financial difficulties, and they had pressure on their water, and they needed to raise the rates, and so the rates kept creeping up every year higher and higher and higher. And finally it got to the point where enough people here in Genesee, Lapeer, Shiawassee Counties, those kinds of places, were upset enough about it that they finally decided, "I think we'd be further ahead to build our own pipeline."

If there were water problems, Hammond said it was just a consequence of aging systems. He insisted that in virtually all cities over seventy years old, "the water infrastructure is that same age or older and is falling apart; needs all kinds of attention. So, we've got leaks all over and not only do we have leaks all over, we have valves that don't work. Whether we stay with the river or whether we were to go back to Detroit, we'd still be losing water out of our system all the time."

Furthermore, he added that Michigan was known for its "impactful Department of Environmental Quality." He insisted that water quality issues weren't of paramount importance—they were being pushed by "professional protesters" coming in from other cities and playing to the cameras. Hammond huffed that "one lady was carrying around a jug. She drew that jug last August, and she's been carrying it around for months. People are under the impression that this is happening right here and now, but her water is perfectly fine now. Another one of those complainers, a new guy, turns out he faked his bottle, and he's admitted this to enough people that the word has gotten out now."

Again, he qualified that if there *were* water problems, well,

> today we're more aware of those problems because we've got so much access to information that we didn't have before. So, people can get worried about stuff, they hop online, they do a little bit of research, and they learn enough to scare themselves. But they don't always learn enough to really have a real answer. And then they start operating out of fear, so instead

of being systematic and saying, "I'm going to get my water tested," people just start screaming and yelling. Well, to me, that is antithetical to the new spirit of Flint.

In the "new spirit of Flint," Hammond urged people to stop causing drama. He told them that they needed to work toward resolution by asking the right questions. Sure, he admitted, the city should have made water test results publicly accessible in a timely manner, but community members needed to trust their public servants. To him, claims that state officials were aware of high levels of TTHM before the public was notified was just "part of the hype."

Hammond was more upset by public arguments and by how some churches have become involved in lawsuits than he was about water quality concerns. "We tried to have some public meetings to get some clarity around this issue and get some questions answered," he recalled. "Unfortunately, the people who were angry and upset wouldn't allow the process to move forward, and then they control the meetings. So, we only had two meetings, and now we haven't had a meeting for more than six or eight weeks."

Hammond decried citizen suspicion. "People are so intent that there is a lot of criminal activity going on," he said. "You can't hardly go to a meeting around Flint where the water issue gets discussed and there aren't also allegations of corruption. I think a lot of it is brought about by fear of the unknown. But, when your water looks clear, and you don't have a problem with it as far as you know, then you don't think about it."

So maybe the city had "fallen down on the job" when it came to notifying residents, but the real problem was that the discussion regarding the water had become so negative that Hammond had difficulty envisioning any successful outcomes. "You don't gain anything by always continuing to think of things negatively," he said. "If you focus on the positive, then you start influencing people to think positively about Flint, about their environment, about what they themselves can do, realizing that we're not going to get answers from the government like we used to get."

Ultimately, Hammond expressed his concern that the water crisis might create new problems. He was passionate about the environment and dismayed by the piles of trash polluting society, and he noted that his fellow Flint residents were switching to bottled water. As he said, "I think that's a huge, enormous ecological disaster that's already happening, and all these water bottles are causing such a huge problem in the environment that is unbelievable." He downplayed the illnesses and the property loss that Flint residents claimed to be experiencing and focused instead on the piles of plastic building up.

Melissa Mays had a vastly different take on the water crisis. A married, working-class mother of three, she had evolved into a formidable advocate for water justice in Flint. I caught Mays at the end of a long workday. She was tired. We were hungry.

Over a meal, Mays shared with me that she and her family hadn't lived in Flint for very long. They were here because of a job transfer. Fresh from David-son, Michigan, Mays and her husband, Michael, were ecstatic about the low cost of housing in Flint. "'Cause [we] lived in the sticks," she added, they were overjoyed to have a Target in town rather than a full hour's drive away.

Like others navigating Flint's working-class and poor areas—and just about anyone who saw Michael Moore's *Roger and Me*—Mays soon realized the city was a shell of its former self. According to Mays, current residents are saddled with the burden of the environmental hazards that corporations like General Motors and Buick left behind. Mays lamented "the fact that GM could get up and leave, not just leaving everybody without jobs but also just leave without addressing the many EPA violations that they have not paid for. They were able to file bankruptcy and get away with not paying the $48 million in cleanup that they were supposed to do."

In some cases, as with GM, the EPA required that companies attempt to remediate land they polluted by planting specific types of trees. But, as Mays recalls, these efforts have not been particularly useful: "Well, they plant [the trees], and then they die. For this tree, its whole purpose is to eat the contami-nation out of the ground. But they have been dying after being planted. You can go down there and just look at the rows of dead stick trees."

During our conversation, it was clear that this environmental justice advo-cate and founder/spokesperson of Water You Fighting For hadn't claimed the job of repairing Flint or protecting its image as an up-and-coming college town. Rather, her creation of Water You Fighting For—which she managed with her husband and the support of her children and close friends—was in response to shifts in the provision of water services and the Mayses' discovery of deleted EPA and City of Flint documents concerning Flint's water distribution system. Through Water You Fighting For and its website, Mays has focused on address-ing the issues concerning Flint's provision of water services and collecting resi-dents' complaints so that they don't slip through the cracks.

Mays and her family hadn't planned to get involved, she remembered, but over time, she'd felt compelled. "In the middle of the day, me and my husband, we went down there [downtown Flint] and then we just joined the protest," she recalled. "It wasn't a march; it was a protest. Then, I started talking to other people who had been sick for a long time and had no reason. People of all ages, races, backgrounds, neighborhoods, were losing clumps of hair, [had] rashes they can't explain." Indeed, shortly after the water source shift in April 2014, Mays and her family began experiencing health problems:

Actually, it started in August. I started to get kind of sick, and then October my nutritionist diagnosed me with Hashimoto's [a thyroid disease]. Hashimoto's? Well, first I'm like, well how did I get this, and she's like, well sometimes these things just happen. So then I got sick again in December, real sick, and I was sick from like the end of December all the way through January, enough to where I had gone to urgent care and they're like, it seems like strep but you're testing negative, so to be on the safe side, we're going to put you on an inhaler. [And] I had a cough that didn't go away and my cough, my phlegm tasted like bleach. We hadn't received the TTHM notices by then. And we didn't know that our water was poison.

The first notice Mays received from the City of Flint regarding TTHM in the water led her to believe that she should connect her poor health to the water source switch. As Mays put it, "We received our first notice in the middle of January. That was the first of them, and nobody knew what it meant. Most people ignored it because it was just a little folded-up piece of paper. But, when I opened it and read it, I was like, you're telling me what now? That it's this unsafe? Like, is this really okay?" Mays mentioned that her doctor and blood tests later confirmed her hunch about the health effects of consuming the public drinking water sold to Flint residents exclusively in Genesee County.

Like most other Flint residents, Mays also mentioned that the first real shift in the provision of water services in Flint affected the water rates. All of a sudden, according to Mays, people in Flint couldn't afford the water. "My bill was like $70 a month; [in recent times] my highest was $765," Mays recalled. "The first month that we switched over, my bill jumped forty bucks. I'm like, I don't understand. And then it just kept going every month, higher and higher and higher. And I'm like, I don't get it." She was especially shocked to see the price of water continue to climb even after her family consciously began consuming less water:

> Right before Christmas, for my shutoff notice, I had to pay a $533 bill. My next bill after that was $765. By this time, obviously, we weren't using the water. Didn't matter. It was telling me on my bill that my consumption was ten units. So I'm like, okay, now you're telling me that I'm using three more units [than during the summer] and that I know that I'm using half the amount of water that I was before?

Faced with the consequences of Flint's recent water shifts and the strong possibility that water quality issues were the root cause of the health challenges that had emerged in her household, Mays made good use of the public relations and marketing skills she'd honed at a privately owned radio station. She felt she needed to get the word out about Flint's dangerous water. She had to do something about it.

Mays insisted that the city had been doing more harm than good by not sharing with residents the pertinent facts regarding the water. "We're being poisoned in our own homes, and when you ask the city about it, they won't answer anything about inhalation or absorption. We've been to so many meetings; we've had so many protests. We've had the EPA here. And he's the one [the EPA representative] who told me that the times that I've had blue bathwater, that's copper." However, city officials, according to Mays, "ignored us." She said, "We've marched, we've protested, we've been on the news, we've been interviewed, we've been on the radio."

As a "water warrior," Mays—along with a multicultural collection of Flint residents including Claire McClinton, who was born in Flint and is from a family of autoworkers, and Flint Rising's Nayyirah Shariff, a community activist—helped residents understand what they were up against in their fight for water justice. Water warriors in Flint also distributed water quality and expense surveys through word of mouth and door-to-door canvassing. And still, the city ignored them. Mays added forcefully, "There's a group of us who are furious. Furious is not even the word. We're furious at the city and the stupid answers we [had] been getting."

When we spoke, Mays seemed incredibly resourceful and resilient in the face of bureaucratic obstruction, though she also shared that some elements of the struggle for water justice bothered her. First, according to Mays, not enough people were involved. Mays and other like-minded folks had done everything they could think of to draw community residents into direct action. They bought ads, printed T-shirts, and distributed some 4,000 door-hanger flyers that Mays described as asking, "Are you having these health issues, is your water like, this, this, this?" They canvassed lots of neighborhoods, trying to cover the bases. But few residents showed up to picket or attend a town meeting about the water. Some never even signed a petition, Mays marveled.

She also mentioned that, while she appreciated the alliance with the Concerned Pastors for Social Action, it was disappointing that so few faith groups were participating. She said that she attended meetings every week with the pastors involved and wondered, "Where's half of your people? Where're all your churches?" According to Mays, "Out of the Concerned Pastors, we basically have like four pastors, five pastors [involved]. There's like 260 churches in Flint." She thought there seemed to be something taboo about this issue; too many church leaders appeared too eager to avoid addressing the water crisis from the pulpit even though it offered them the biggest spotlight and a chance to make a significant impact every Sunday morning:

Concerned pastors have opened their doors to us. The problem is that the other churches that we've emailed, called, stopped in [at], want nothing to do with it. It's almost like tackling the homosexual issue in a Catholic church; that's how I feel. If I was a lesbian and said, "I want to come into your church and talk about how you should treat homosexuals fairly," that's the kind of blowback we're getting. No, "I don't want to talk about it; it's too controversial," [they say]. Explain that to me! I feel like I'm dumber, not even from the water, just from people's responses or lack thereof. I just don't understand it. Even my kids go down to protests with me, rain, sleet, snow. They're ten, twelve, and sixteen and they know it's wrong.

State leaders, according to Mays, claimed that the water couldn't *really* be a problem if more Flint residents weren't joining efforts to fix it. "Nobody must be that upset about it if there are not thousands of people out marching," Mays mimicked. "I was like, it's a struggle for fifty people to show up on a nice afternoon," she said, musing how much harder it was to convince even a few people to march in the middle of a Michigan winter or to stop by to join the picket after a long day at work. But without residents' consistent agitation, leaders in the fight for clean, affordable water were having trouble showing the city and the state that they were a force to be reckoned with.

For instance, when it came to water testing, Mays said that, without the impact of loud protests on behalf of better water, the city was "only letting one person go out and test these houses, and that's the lab supervisor who should be in the water plant, running the water plant. He's only allowed to test for, like, TTHMs. He's also limited in the number of tests he can do per day. That's six tests a day."

It wasn't even a drop in the bucket. Some residents secured water test kits directly from the water department and attempted to test their own water. However, according to Mays, the water department claimed that few people ever brought back the specimens. The only other option for having their water tested was private testing (for about $400) or asking the DEQ to test for TTHM or lead and copper (again, at the resident's own expense). So, Mays was frustrated. Taking to the streets seemed to be the only way to convince people that their water was poison, and no one seemed to want to take up her call or pick up a picket sign.

Even those who agreed the water was contaminated were divided, Mays said with regret. She was concerned that people in Flint were starting to become unhinged, appearing inarticulate and angry to outsiders unfamiliar with the back-and-forth between water rights activists and local and state officials. No matter how it made Flint residents look, without fail, someone would eventually lose his or her patience during the meetings with the city. "People [would get] up yelling," Mays recalled. "They make us look like crazy animals because we are yelling 'cause they're not answering our questions." In response, when advocating for clean and affordable water, Mays maintained that she tried to "behave [her]self in meetings."

Mays mentioned that she "always wanted to be the calm face." As she put it, "I will ask questions, but I will not yell, and I will not interrupt. Even though I want to tear things up, I don't want to be labeled as crazy or rude or whatever. I organize the marches. I do a lot of this, but I'm too nice for them to hate. I'm pissed, but I'm not going to let them discredit all the things I'm saying by calling me crazy."

Mays went on to voice frustration with the fact that some residents seemed to have their own agendas. She hinted that these included folks like Flint City councilmen Scott Kincaid and Eric Mays, who were running for mayor and seemed to be using the water issue to catapult their election bids.

Mays also complained about attempts to make the water issues "about race." She was put off by individuals who focused not on *everyone's* right to clean and affordable water but on the racial disparities in impacts and environmental justice. She recounted one person telling her that the water issues mattered more to Black residents, repeating the theory that the areas most affected by the water crisis were predominantly Black American. But Mays felt that the water problems weren't limited to racial minorities. "I'm like, dude! The water isn't saying, 'Oh, you're this race. I'll poison you more than the other.' I was like, no, that's not the way I want to go with it. I want to go with it in an educated and passionate way. Throwing out the race card makes people leave. The poor people, the regular people—that's what I say, the regular people—are getting destroyed."

To her, polluted water was hurting everyone, and bringing race into the discussion was just another distraction: "We're all getting poisoned. We have jobs; we pay our taxes, we pay our water bills, we're paying for our own poison. We're paying to have our children poisoned in front of us. All of us."

Finally, Mays pointed out that water quality issues could be found in many urban and rural communities throughout the United States. Flint is a nonfactor to the outside world, she told me, and it would take a significant push to get attention and get the water fixed:

We're not Ferguson, and we're not Baltimore. So, nobody cares. That's the biggest feeling overall: nobody cares what happens to Flint. We have such a bad reputation as it is, nobody cares. The other thing is that all of us are too broke; we can't destroy our other broke friend's stuff, so what's the point? We talked about having sit-ins. But none of us can afford to go to jail. Plus, I would stand to lose my job, my health insurance to pay for the little it's paying for. So it's like, no I can't. I would love to, but you know what, no, I can't. Most of us are way too broke to lose what we have.

Bill Hammond's and Melissa Mays's worlds were not separate. Yet the divide in their stances on the *same* issue were vast. When we spoke in May 2015, Hammond believed the water crisis was a passing issue that frustrated people were using to get riled up about problems they didn't really understand and to advocate for solutions that would not begin to address the long-standing concerns that poor and depopulating cities like Flint faced. Mays believed the water crisis was a clear and present danger, infecting the lives of all Flint, including the residents, businesses, and churches, and those who opted to ignore this crisis were doing so at their own peril.

To me, it looked more and more like Mays and Hammond were both, to some degree, right. As evidence mounted that the water issue could not be ignored, change would be harder to achieve, because Flint had already been devastated by neglect and distracted by neoliberal plans to rebuild. Hammond's structural concerns touched on the conditions for all that ailed Flint—and, without addressing them, for all sorts of new problems that would surely arise—while Mays's immediate water crisis was a critical symptom that could no longer be ignored.

PART II

After Evidence of Lead
Contamination Surfaced
in Flint's Water

Well before Flint residents levied numerous complaints about the quality of their water and obligation to pay for a public resource that they were scared to use, essential services in this discarded city had been on the decline for decades. Federal efforts like the Model Cities program, in addition to local initiatives—including the Better Environment for Neighborhoods program and plans to revitalize the city's downtown area—lacked the capacity to drive fundamental changes within Flint, especially in its struggling neighborhoods. It is also worth noting that the segregationist roots of Flint's political structure had proved difficult to uproot and that access to quality schools and community resources, particularly within poor areas in the city, had grown worse over time.

Michigan's policies over the previous decades had not improved conditions in its low-resource municipalities. The state's cities had been defunded to the point of breakdown, enabling the state to introduce nonelected administrators to oversee local governments and to wield broad powers and authority to auction off city-owned assets, demolish union contracts, and privatize everything that was not nailed down. Michigan's fiscal policies helped put its aging cities in the situations they now faced.

Making matters worse, with pockets of the Flint community fearing the long-term effects of consuming water supplied by the city and other residents claiming that issues with it were temporary, Flint couldn't be more divided. Despite the steady parade of residents picking up cases of water from local giveaways, some people continued to question just how bad the issues were in Flint, maintaining that problems with the water were overblown and complaints of citywide contamination were fabrications. However, these naysayers did not silence the complaints of families who regularly drew cloudy water from their faucets, something that became common after the financially strapped city cut ties with the Detroit water system and began drawing directly from the Flint River under the "leadership" of state-appointed emergency managers. Many Flint residents, especially those experiencing health problems after the water source switch, felt that their lives and well-being were being compromised.

All the while, the City of Flint—as well as the Michigan Department of Environmental Quality—continued to claim that the water was safe to drink. Somehow, officials expected residents to believe that water with chloride levels so high that General Motors refused to use it in its West Bristol Road engine plant was safe for human consumption. When residents complained about discolored "rusty" water, the city blamed the miles of aged pipes that occasionally released iron into the drinking water. Residents who were concerned about overall water quality were told to flush their tap for 30 to 120 seconds before drinking or cooking with it. The city also provided reassurances by claiming that the appropriate industry professionals and state regulators were on board with the monitoring of ongoing water issues.

State and local officials, as well as the water regulators, continued to push back on claims indicating significant water contamination issues in Flint. But there were federal EPA protocols that could have been implemented to mitigate or remediate water contamination threats—and decrease the impact on the public—when concerns about drinking water contamination surfaced. For instance, the EPA has a blueprint—as specified in its "Response Protocol Toolbox"— on how water utilities should plan for, and respond to, drinking water contamination complaints. Since all drinking water systems are vulnerable to contamination, water utilities contending with a credible contamination threat are advised to follow specific steps, including confirming the credibility of the contamination threat, identifying likely consequences of water contamination, and determining how responses will impact the public. Due to the potentially catastrophic consequences of failing to implement a timely contamination threat response, the EPA recommends that water authorities and state regulators figure out whether the threat is credible in a relatively short amount of time. After evaluating data that tracks water quality to help identify abnormalities across time and space, deviations from norms or the baseline are supposed to be investigated to determine the legitimacy and scope of the threat.

For example, consumer complaints about the smell and taste of water—as well as health problems attributed to it, like skin irritation—should, according to EPA protocols, prompt an investigation aimed at confirming the threat that is causing havoc among exposed populations. Water treatment professionals and regulators, triggered by persistent and consistent complaints about the taste and odor of water, should first attempt to examine how complaints vary by type and location to determine whether the aesthetic qualities of the drinking water have been caused by contamination. Water utilities and regulators are encouraged to take consumer complaints seriously. Training for the staff in fielding complaints from the public is also urged so that protocols to deal with unfamiliar trends in issues are in place to report, investigate, and address these claims.

Once the threat is evaluated and confirmed, the EPA recommends several tasks in the threat management process that should be executed to protect the public from contamination, such as site characterization and an operational response or effort to mitigate the contamination. For instance, not only are public water monitors and regulators expected to identify the scope of the problem within the water system, but also appropriate threat responses that involve efforts to keep consumers away from the contaminated water by issuing water warnings and providing a temporary water source until issues are resolved. Relatedly, operational containment strategies should include an analysis that identifies the potential health effects of the contamination. Finally, the remediation process should incorporate a variety of tasks and activities geared toward identifying available options to remedy the contamination and additional ways of monitoring the water system to ensure the threat has been eliminated.

The truth of the matter is that, while it is impossible to eliminate all sources of potential contamination in water systems, utilities are capable of developing and executing plans for responding to unusual threats when they emerge. The City of Flint's initial response to water quality concerns was delayed and heavily undermined by contradictory actions and statements. What we shall see in the chapters that follow is that Flint's approach to resolving this problem, particularly after the water contamination threat was confirmed, differed vastly from the protocols outlined by the EPA.

And the fact is that Flint didn't have to go it alone. At the national level, there was expertise and experience that the federal government could have brought to bear to aid the city's residents. Likewise, the state of Michigan had a vast store of institutional knowledge about water quality. But instead, Flint's community leaders were often left in the position of having to plot their own course—at times with the aid of experts who were already accustomed to fighting federal and state bureaucracies—while officials on the higher rungs of government either looked the other way or actively hindered efforts to clean up Flint's water.

5

Lights, Cameras, Interventions

The Flint Crisis in the National Spotlight

Getting the word out about Flint's hazardous, foul-smelling, and discolored water took tenacity. Sure, some Flint residents thought the water was fine. But others, including Melissa Mays, diligently worked to bring attention—and change—to Flint. Local water activists, hoping to secure proof of their concerns about the water, reached out to Miguel Del Toral, the EPA Region 5 Ground Water and Drinking Water regulations manager. He is the author of the June 2015 EPA memo leaked by the ACLU on July 9, 2015, and the corresponding emails sent to higher-level administrative staff at the EPA Region 5 headquarters. He began voicing his concerns in February 2015 regarding elevated lead levels found in Lee-Anne Walters's Flint home.[1]

Del Toral had been pleading with his superiors at the EPA to take seriously his research on the city's water—research that he had initiated due to his concern for Flint residents. He even asked for permission to use his own money to do more water testing in Flint. But his supervisors at the EPA failed to heed his persistent warnings about high lead levels and the deceptive actions by City of Flint water treatment professionals that had helped cover up the presence of lead in the water—actions that he believed "bordered on criminal neglect." He also faulted the city for several other things, including its failure to notify Flint residents about the lead problem.

Del Toral would later go on to describe the US EPA as "a cesspool" because of the agency's reluctance to intervene regarding the high lead levels in Flint's drinking water. He admitted that he "truly, truly hate[s] working here." Explaining this frustration, Del Toral wrote in June 2015 that "it's all of this 'don't find anything bad' crap at EPA that is (the) reason I desperately want to leave.... I am not happy to find bad things.... It is completely stressful because it means children are being damaged and I have to put up with all of the political crap, but where these problems exist, I will not ignore them."[2]

So, Del Toral suggested that Flint residents enlist the help of Virginia Tech civil engineer Marc Edwards to test the water. Dr. Edwards is known as the "troublemaker" scientist[3] and has been a vocal critic of the EPA's permissive regulatory culture for years, calling the agency's Lead and Copper Rule "a national embarrassment."[4] He agreed to take on the task of testing Flint's water.

After securing emergency funding from the National Science Foundation, Edwards headed to Flint with a small group of graduate students to support his efforts.[5] He also enlisted help from Flint residents. Lee-Anne Walters, also known as "Resident Zero,"[6] helmed a team that included fellow Flint residents Melissa Mays, Claire McClinton, and Nayyirah Shariff, alongside community churches and organizations such as the ACLU. Together they spread the word about water testing efforts. They also planned and coordinated the distribution of water testing kits and ensured that kits were delivered to residents who had transportation challenges.[7]

The response rate for testing efforts was extremely high: 92 percent (277 of 300) of tests were returned in August 2015, 68 percent (184 of 269) in March 2016, and 91 percent (164 of 180) in August 2017. In the end, 156 Flint residents assisted water sampling campaigns throughout all testing phases between August 2015 and November 2016.[8]

Water testing provided the needed answers. Flint's citizens finally had hard evidence: almost 17 percent of the samples indicated lead levels at 15 parts per billion (ppb) or above. At that level, the city was legally required to initiate a public intervention.[9] The city, of course, claimed that samples had been collected only in the oldest parts of Flint, while Virginia Tech researchers countered that, in their initial rounds of testing, high lead results were present even in homes with lead-free plumbing. Their data suggested that city-owned lead pipes and corrosive water marked the real source of the crisis in Flint.[10] In many cases, lead compliance testing results generated by Virginia Tech were higher than the results for the same home supplied by the Department of Environmental Quality.

Medical professionals added their own data to the push for change. In September 2015, Hurley Medical Center doctors, led by pediatrician Mona Hanna-Attisha, concerned about unexplained health issues she observed in Flint's children after the shift from Detroit water to Flint River water,[11] published a study indicating a spike in the lead levels of Flint children after the April 2014 switch. Hurley Medical Center's analysis relied on data drawn from all blood lead levels (BLLs) processed by the Flint area hospital. On average, Dr. Hanna-Attisha's team found that the percentage of children with elevated

BLLs doubled after the shift and that levels differed significantly by ward and zip code. Children residing in zip codes 48503 and 48504 had the highest BLLs in the sample, with 2.5 percent having elevated BLLs before the switch and 6.3 percent having elevated levels after.[12]

In addition to bringing attention to the recent uptick in BLLs, Hanna-Attisha's analysis revealed a link between the elevated BLLs observed and the uneven distribution of the water problem across Flint. Her team verified that there was a connection between the rise in BLLs among children and the patterns of lead-contaminated water that Flint citizens and Dr. Edwards's research team had discovered during their testing in 2015. High water lead levels found disproportionately in homes in some wards (such as Wards 5 and 6) than in others (such as Ward 9) were associated with elevated BLLs among Flint's children during the post-switch stage.

Two weeks after the public learned about Dr. Edwards's findings that Flint's water was nineteen times more contaminated than Detroit's water,[13] the results of Dr. Hanna-Attisha's study of lead exposure among Flint's preschool children became the catalyst for change in Flint. News of the combined research efforts spread by press conferences, media reports, and word of mouth. While Flint residents gave media interviews and provided updates via Internet posts in order to bring attention to water issues and motivate corrective action, Michigan's DEQ was taking advantage of its head start, given that it had been aware of these findings well before they were presented to the public. As the DEQ's public information officer Karen Tommasulo put it after the ACLU ran a story about Flint on July 9, 2015, "Apparently it's going to be a thing now."[14]

Brad Wurfel, the DEQ's communications director, tried to help recast public perception about the water crisis. For instance, when Del Toral's memo was exposed, Wurfel characterized Lee-Anne Walters's water issue as an "outlier" and directed anyone with a home older than thirty years to get their water checked, just to be safe. But in general, he mentioned that everyone else should put the concern to rest. As he put it, "Anyone who is concerned about lead in the drinking water in Flint can relax."[15]

Later in July 2015, rather than claim there was no problem with lead in Flint water, officials like Stephen Busch, then district supervisor of the DEQ's Office of Drinking Water and Municipal Assistance, simply claimed that neither Flint nor the DEQ was at fault for the water quality concerns. According to Busch, "[Lead] is not coming from the Flint River or the City's Water Treatment Plant or the public distribution system. It is from lead service lines into homes and from plumbing materials and fixtures within the private property of the household."[16] There was a problem, the DEQ now seemed to confirm, but individual homeowners and their aging plumbing were to blame.

The DEQ went further. It publicly blamed community groups for fanning the flames of fear in Flint. Wurfel wrote in an email to Michigan senior federal policy representative Eric Brown that "folks in Flint are upset because they pay a ton for water and many of them don't trust the water they're getting, and they're confused, in no small part because various groups have worked hard at keeping them confused and upset. We get it. The state is trying like mad [to] get the word out that we're working on every aspect of the health safety of local water that we can manage, and the system needs a lot of work."[17]

The DEQ also attempted to undermine the Virginia Tech water study.[18] Wurfel claimed that residents concerned about lead in drinking water should consult a certified laboratory, and he accused the Virginia Tech researchers of biasing results by sampling only the city's oldest homes. As he noted in a September 9, 2015, email to reporter Ron Fonger, "This group specializes in looking for high lead problems. They pull that rabbit out of that hat everywhere they go. Nobody should be surprised when the rabbit comes out of the hat, even if they can't figure out how it is done. . . . Quick testing could be seen as fanning political flames irresponsibly. Residents of Flint concerned about the health of their community don't need more of that."[19]

By the second week of September 2015, residents' concerns about lead continued to mount, and a resolution to the water situation couldn't come soon enough. This was around the same time that the public alliance between Flint and the DEQ began to show wear and tear.

Earlier in the year, on July 9, Mayor Dayne Walling was seen on Flint's WNEM TV5 news consuming a cup of Flint drinking water and assuring residents, "It's your standard tap water." He added that "you can taste a little bit of the chlorine."[20] But by September 11, 2015, Walling was making public attempts on his reelection website to distance the city's actions from the DEQ's with the statement, "The City will be continuing to optimize its water treatment process including planning to use a corrosion inhibitor now that it is being allowed by the DEQ."[21]

The DEQ's Stephen Busch told Brad Wurfel that the city had not communicated its intentions to the DEQ and that the regulatory agency, in its August 17 correspondence, recommended the city start using the corrosion inhibitor phosphate.[22] Liane Shekter-Smith, with the DEQ's Office of Drinking Water and Municipal Assistance, added, "It should be noted that the city does need to obtain a construction permit to install treatment. They have not yet applied for such a permit. So I'm not sure what the mayor means about us finally allowing them to proceed. The ball's in their court."[23]

A little over a month after Dr. Hanna-Attisha shared her findings with the public on September 24,[24] tensions between local leaders and state regulators escalated. Mayor Walling went further in an interview posted online on October 29, 2015, in which he implicated the DEQ in the mismanagement of Flint's water system. Had the state provided oversight and adequately helped Flint transition from the Detroit water system, Walling contended, the city would have been able to protect residents from tainted water. Technical guidance and monitoring from the DEQ, he claimed, would have improved the city's efforts to protect residents from environmental hazards. "Authority to permit water treatment for drinking water is held by the Michigan Department of Environmental Quality," he stated. He went on to say, "Whether the City staff discussed it five times or 50 times or 100 times, the authority for approving it is with the State. . . . The City relied on assurances from the Department of Environmental Quality that we were meeting the standards, we were complying with the Safe Drinking Water Act and Lead and Copper Rule when, in fact, it wasn't."[25]

Despite the intense exchanges between local leaders and state regulators, after the water issues were confirmed by independent researchers and reported to the public, state and local administrators seemed to leap into action by launching efforts to provide safe water and requesting state and federal assistance.[26] On October 2, 2015, state officials and Mayor Walling promised Flint residents the water crisis would soon be resolved. Two weeks later, on October 16, the Michigan legislature approved $9.3 million to help Flint switch back to corrosion-treated water from the Detroit water system. Funds were set aside to provide free water filters to citizens, increase the number of public health workers, and test children in Flint schools.[27] And in November, just over 18 percent of the 75,000 registered voters in Flint chose to replace Mayor Walling, a previously favored local leader whom residents no longer trusted, with a local psychologist, Dr. Karen Weaver.[28]

Still, after Flint returned to Detroit water on October 16, 2015, the water remained unsafe to drink. Flint's prolonged failure to treat corrosion had damaged the coating inside pipes and rendered the water crisis not easily reversible.[29] Despite this reality, city and state officials were in the hot seat. By early November 2015, they were spending a great deal of energy attempting to deflect blame by trying to placate the masses with promises to rebuild and reignite that spirit that Bill Hammond spoke about. However, as I show in this chapter, public health and water quality administrators weren't incredibly forthcoming with what they knew. In addition to being too slow to respond, public health administrators at this stage were instrumental in initiating Flint's second shift, now toward malign neglect.[30]

Customers' Frustration with High Water Costs

People openly voiced their frustration with decision makers' actions once they knew that local, state, and regional water regulators had been aware of the elevated lead in Flint water before residents proved that it existed. Regulators had failed to require corrosion control and delayed efforts to protect public drinking water from contamination after over a year of water equity protests by residents. And those regulators were taking the heat for it.

At city council meetings, town meetings, and panel discussions about the water crisis, Flint residents were especially focused on water rates. Keith Pemberton, an elderly white gentleman who had been actively involved in protests concerning water cost and quality since January 2015,[31] was adamant at a November 9, 2015, meeting that the citizens of Flint deserved a break from paying for contaminated water. He spoke passionately about the medical community's need to be concerned about Flint residents over the age of five, given that they "have been poisoned with lead [and] aren't in the group that's to be treated." But his main concern that evening was the cost of water:

> The water bill . . . this isn't a laughing matter. People have been paying the highest prices in the nation, which you already know. They're paying for poison water and you are requiring them to continue to pay for poison water. I have the perfect way for you to get the bills paid: Give us the refunds we're supposed to have. Add that to the bill. I can't believe that my city, *my city*, is going to be screwed this Christmas! And in December, [the city plans to] turn off people's water? I just can't believe that. But you know what? It's going to happen because of the heartless, gutless people that we have making decisions.[32]

Other residents, like Carolyn Shannon, a Black woman who attends city council meetings often, also blamed the slow response on the fact that Flint remained under state receivership and on the inevitability that "someone making the rules for us will not work." She mentioned at this same city council meeting that she was appalled by Flint being managed by outsiders who run "this city on the backs of senior citizens and the less fortunate." Now, with national attention on the water quality issues, Ms. Shannon seemed to think that problems with water quality would eventually be resolved.[33]

During this time, many Flint residents who had not been paying their water bills were starting to fear that their water would soon be turned off—this despite the August 2015 legal victory in a civil case filed by two Flint residents in 2014 that required the city to halt water shutoffs and make adjustments to delinquent water bills until they rolled back the 35 percent rate hike of 2011 deemed illegal by Genesee County judge Archie Hayman.[34]

Residents at this November 9, 2015, meeting were concerned that they had not seen any credits and that their accounts were still considered delinquent. Since her present concern was the city's water rates, Ms. Shannon used her time at the mic during the public comment period to advocate for people who could not afford to pay for the water that they had recently learned was contaminated with lead. According to her, "Those who have jobs, they can pay their water bill. I am asking you not to turn our people's water off. Do not send them any shutoff notices. . . . I am my brother's keeper and we should all be trying to help with this water situation. . . . Clean up the water. Send us our refund in the mail before Christmas. That's what you do if you love the City of Flint. . . . Send us our refunds."

Unfortunately, water credits did not emerge by Christmas. Instead of refunding the money that residents had overpaid, the City of Flint appealed Judge Hayman's August ruling. In December 2015 the city convinced him to revisit the preliminary injunction that had forced the city to lower its water rates and refrain from sending shutoff notices—both of which were harming Flint's financial stability. The city claimed that the court-ordered lower rates that it had been charging since September had cost them about $3.4 million. In response to the city's petition, Judge Hayman granted permission to issue shutoff notices to customers with delinquent water bills. Doing so opened the door to disconnecting accounts belonging to people who had been late paying their water bills since September 2015 and to those customers who had had an overdue balance before July 2012. Consequently, instead of receiving water credits, thousands of Flint residents received notices announcing that their water services would soon be disconnected.[35]

Deceitful LCR Testing Procedures

As residents noted, local officials and regulators certainly had enough blame to divvy up and pass around, especially given that plenty of people were afraid that their water would be disconnected due to nonpayment—after they had been informed that the water was unsafe to consume.

Public administrators and regulators involved in the crisis also received plenty of criticism when some of their decisions—which prolonged the crisis— came to light, including the city's routine rigging of the 1991 Lead and Copper Rule (LCR) compliance testing. Correspondence between DEQ official Adam Rosenthal and Flint utilities director Mike Glasgow, uncovered by FOIA requests made by Dr. Marc Edwards, reveals a focus on homes expected to have low water lead levels and conveys the officials' intentions to manipulate LCR test results—even though state and local regulators were aware of potential lead problems in the area.

"We hope that you have 61 more lead/copper samples collected and sent to the lab by 6/30/15 and that they will be below the AL [action level] for the lead," wrote Rosenthal. "As of now with 39 results, Flint's 90th percentile is over the AL . . . for lead." The LCR requires that water suppliers test high-risk homes. But Flint officials approached residents of Flushing Road to test their water, confident that these residences would show low water lead levels. Their confidence originated from the fact that the city had replaced a section of the Flushing Avenue water main in 2007, and mains that have been replaced are far less likely to have water quality issues.[36]

Miguel Del Toral's leaked EPA Region 5 memo shows the extent to which state-level regulatory and city officials disregarded LCR policy and the evidence of high water lead levels in Flint. Del Toral, the Ground Water and Drinking Water Branch regulations manager, wrote to Thomas Poy, the branch chief, summarizing, "A major concern from a public health standpoint is the absence of corrosion control treatment in the City of Flint for mitigating lead and copper levels in the drinking water." The LCR requires systems serving 50,000 or more people to implement and maintain corrosion control treatment for lead and copper, but as of June 24, 2015, Flint was not following protocol. Before the switch to Flint River water, the city had purchased treated water containing orthophosphate, a chemical that controls lead and copper levels in drinking water; after the switch, Flint stopped orthophosphate treatments.[37]

Considering Flint's lead problem, Del Toral advised the EPA to follow up with Flint and the DEQ. However, the agency failed to comply with Del Toral's suggestions and did not appear to share his concerns about the city. Correspondence between the EPA Region 5 and DEQ officials before and after the public's discovery of high water lead levels shows that Del Toral's recommendations were ignored. Between February and June 2015, Del Toral's concerns were dismissed despite the regulatory agency's prior knowledge of the city's failure to treat corrosion.[38]

In an April 22, 2015, correspondence with Poy, Del Toral advocated for Lee-Anne Walters, who was in danger of having her water shut off and being evicted unless she dropped claims against the city for lead poisoning in her children. Del Toral characterized the DEQ's stance toward the claim as "completely inappropriate (at best)." He informed Poy that "on March 23, 2015, Steve Busch (DEQ) left me a voicemail . . . indicating that this was the resident's problem . . . that she needed to hire a plumber to fix it and that DEQ was not planning any further action." Del Toral contended that responsibility for addressing the problem should not be delegated to the water customer: "The city should be required to remove the lead hazard posed by their lead line and restore water service to the home as it is completely unreasonable to shut off someone's water for something that appears to be the fault of the city."[39]

Del Toral stood up for Flint through his escalating concern for Lee-Anne Walters and her family. Without his persistent advocacy and Dr. Edwards's FOIA requests that unearthed conversations between water administrators, there would be little evidence demonstrating just how far water regulators and administrators had gone in their efforts to turn a blind eye to Flint's water problems. Had Del Toral not gone beyond the call of duty to consistently warn water authorities about the lead problem in Flint, the city would likely not even know that the water was tainted with lead to this day.

Del Toral repetitively expressed concerns over high lead levels in Flint. Plus, he raised the alarm for residents because corrosion control—an effort to prevent pipe deterioration by adjusting the water's characteristics, including its pH and alkalinity—had not been implemented.[40] In addition to these problems, he and Edwards both raised serious concerns with LCR testing procedures. Before the water crisis, the City of Flint Water Treatment Plant directions for testing residential drinking water for lead and copper included instructions to run the water for three to four minutes before sampling.[41] Similar requests to preflush pipes prior to testing water were provided to Michigan residents in Grand Rapids, Detroit, Muskegon, Jackson, and Holland between 2007 and 2015.[42]

It is important to note that this practice is not uncommon in the United States. In cities like Philadelphia, as recently as 2014 water departments asked residents to run their water several minutes before testing.[43] Although the EPA is aware of the widespread use of pre-flushing during LCR compliance testing and has acknowledged that pre-flushing the pipes is harmful to the testing process because it biases results, the agency has not exercised its authority to force municipalities to abandon the use of this pretesting method. The EPA is not required by the LCR to approve or disapprove local municipalities' sampling protocols, despite pleas from residents in cities like Washington, DC, to address this issue.[44] In other words, by law, all the EPA has to do is require that municipalities do the compliance testing. But, this agency does not have the capacity, or has not exercised its authority, to make sure that the tests are done correctly.

This inaction did not stop Miguel Del Toral from warning water authorities about the use of pre-flushing in Flint. In a February 27, 2015, email addressed to various US EPA and DEQ officials, including the EPA Region 5 Michigan program manager Jennifer Crooks and DEQ district engineer Michael Prysby,

Del Toral warned that Flint's method of preparing compliance samples for lead testing—instructing that taps be flushed for several minutes before a sample was taken—biased the results. As he put it, "People are exposed to the particulate lead on a daily basis, but the particulate lead is being flushed away before collecting compliance samples." He went on to say that this method of testing provides "false assurance to residents about the true lead levels in the water."[45]

It turns out that Del Toral had good reason to be concerned about the use of pre-flushing procedures in Flint. Children weren't being exposed to lead contamination only at home. Data from recent LCR compliance testing within Flint Community Schools facilities show that many of the taps, including drinking water fountains, classroom sinks, and kitchen faucets, contained lead-contaminated drinking water. And testing showed how much of this lead would have been undetected had the DEQ not agreed to suspend water testing norms at the urgings of Del Toral and Edwards.

Table 5.1 documents the percentage by school of taps within Flint Community Schools facilities with lead levels over 15 parts per billion (ppb) between late 2015 and January 2016. The first draw, which did not involve pre-flushing the taps, showed extraordinarily high evidence of lead in school drinking water. While just 2 percent of the taps in the Michigan School for the Deaf contained water above the actionable level for lead, the percentage of taps contaminated in the Brownell, Doyle-Ryder, and Pierce Elementary Schools was 53 percent, 40 percent, and 74 percent, respectively. Taps in Flint's high schools ran with lead and copper, too.

It's easy to see from table 5.1 that detectable lead levels are sensitive to sampling procedures. With the exception of Flint Northwestern High School's data, the detectable amount of lead in drinking water declined significantly after a thirty second pre-flushing and became undetectable in most of the schools after a two-minute pre-flushing. And make no mistake: pre-flushing can temporarily reduce the amount of lead a water user consumes, but it is exceedingly unlikely that children at school were flushing taps for two minutes before drinking the water or washing their hands in it.

Each facility's report included multiple references to replacing the plumbing in impacted faucets with lead-free materials. That was because over half of the taps that did not have lead-contaminated water contained plumbing materials similar to those taps in which actionable lead levels were detected.

Inspection teams urged Flint Community Schools to replace the plumbing in nearly *all* of the taps in its facilities, not only those where the tests revealed elevated lead levels. This is an expectation that Flint Community Schools should have met years before due to the requirements of the Lead Contamination

Table 5.1. Percentage of Flint schools with actionable lead exposure after pre-flushing, 2016

	Percentage Taps over 15 ppb		
School/Facility	No Flushing	30 Sec. flushing	2 Min. Flushing
Brownell Stem Academy	53	4	0
Doyle-Ryder Elementary School	40	2	0
Durant Tuuri Mott Elementary School	43	10	3
Eagle's Nest Academy	60	0	0
Eisenhower Elementary School	42	0	0
Flint Schools Central Kitchen	27	9	0
Freeman Elementary School	29	3	0
Holmes Stem Academy	35	2	0
Manley School	32	2	0
Michigan School for the Deaf	2	0	0
Neithercut Elementary School	46	0	0
Northwestern High School	64	32	23
Pierce Elementary School	74	18	8
Potter Elementary School	35	7	0
Southwestern Classical Academy	69	18	4

Sources: Brownell Stem Academy: Outlet Sampling and Plumbing Assessment Recommendations; Doyle-Ryder Elementary School: Outlet Sampling and Plumbing Assessment Recommendations; Durant Tuuri Moot Elementary School: Fixture Sampling and Plumbing Assessment Results; Eagle's Nest Academy: Fixture Sampling and Plumbing Assessment Results; Eisenhower Elementary School: Outlet Sampling and Plumbing Assessment Recommendations; Flint Schools Central Kitchen: Fixture Sampling and Plumbing Assessment Results; Freeman Elementary School: Outlet Sampling and Plumbing Assessment Recommendations; Holmes Stem Academy: Outlet Sampling and Plumbing Assessment Recommendations; Manley School: Fixture Sampling and Plumbing Assessment Results; Michigan School for the Deaf and Learning Resource Center: Fixture Sampling & Plumbing Assessment Results; Neithercut Elementary School: Fixture Sampling and Plumbing Assessment Results; Northwestern High School: Fixture Sampling and Plumbing Assessment Results; Pierce Elementary School: Outlet Sampling and Plumbing Assessment Recommendations; Potter Elementary School: Outlet Sampling and Plumbing Assessment Recommendations; Southwestern Classical Academy: Fixture Sampling and Plumbing Assessment Results. School data: https://www.michigan.gov/flintwater/0,6092,7-345-76292_76294_76297---,00.html.

Note: pbb = parts per billion

Control Act of 1988, which requires schools to identify and remove lead-bearing plumbing materials from all water fountains.[46]

Despite all the blame being tossed around, there was surprisingly limited backlash to Flint residents' prolonged exposure to toxic conditions—exposure that stemmed from years of benign neglect, disregard for necessary repairs to the public infrastructure, and the provision of public services that took hold during the Nixon administration.[47] Funding set aside for public improvement and land maintenance declined substantially during this time, while most community development block grant funds were consistently allocated to ongoing urban renewal projects in the city.

Investments in Flint's public infrastructure continued a downward slide during the 1990s. Why? Downtown revitalization efforts failed and economic restructuring led to gradual population changes that contributed to significant declines in property tax revenues, shifts in job availability, and further community impoverishment. Periods of financial insolvency between 2002 and 2006, and then between 2011 and 2015, further diminished the revenue and property values in Flint.

When developments like this happen, a city's ability to provide public services in the short and long term is severely undermined. So, it is no shock to find that Flint could no longer cover the resources or the manpower necessary to provide public services effectively. Depopulated cities like Flint, with fewer people paying taxes and city fees, can't afford aging workers and their pensions, side street cleanup, road resurfacing, or water main and meter repairs.[48]

People were distracted by the costly decisions made by water administrators and regulators—decisions that were buried in the hundreds of pages of correspondence secured by Dr. Edwards—and by the constant competition for higher moral ground among agencies. So there wasn't much speculation about what health and toxic exposure data alone should have revealed to water regulators and administrators. There also was not much inquiry into how the improper deployment of this monitored health data prolonged the water crisis—and crippled the city's capacity to implement a targeted response to a public health event that had disproportionately hit specific areas in the city.

Water authorities and local municipal and health officials had access to data from Michigan's Lead Poisoning Prevention Program that monitors blood lead levels in Flint and other Michigan cities. But they failed to intervene concerning observed lead exposure spikes in this community, opting to do the least. They also neglected to apply available research that could have guided efforts to locate the populations most impacted by the crisis. Instead, they just watched while health providers in Flint consistently overlooked their obligation to retest BLLs in children who had elevated results that indicated lead poisoning.

Overlooking the Relationship between
Race and Lead Exposure in Flint

Dr. Mona Hanna-Attisha's research team had released game-changing results in September 2015. And if health authorities who received BLL test results for all children within seven days, on average, had been trying to aid the situation, they could have suggested efforts to look beyond ward differences and examine BLL variations at the census tract, if not the parcel level.

Children's BLLs varied considerably across census tracts. Most census tracts were close to or slightly higher than the average BLL among preschool children in Flint (2.59 µg/dL). Some census tracts (such as 1, 5, and 34) contained children with lower BLLs than the city's average during the water crisis (2.49 µg/dL). Meanwhile, other census tracts (such as 10 at 3.94 µg/dL, 15 at 4.03 µg/dL, and 29 at 4.44 µg/dL) were significantly higher than the norm during this time.

Health regulators also overlooked demographic patterns that varied from trends in recent years. While Hanna-Attisha's analysis was helpful in explaining BLL differences among children by ward and zip code, it did not provide much insight into how these patterns varied across race and income groups. Such insight would have been helpful, given Flint's history of residential segregation and the back-and-forth among residents regarding the significance of race in causing the water crisis.

Before the water source switch, there was every indication that gaps in lead exposure among racial/ethnic groups were in decline. For instance, I learned from an independent t-test (a procedure that determines whether the difference between two means is statistically significant) that by 2010, blood lead exposure gaps between racial groups among preschool children in Flint were minimal. This analysis of Michigan Department of Health and Human Services data showed that mean BLL differences between Black (M = 2.32, STD = 1.957) and white (M = 2.25, STD = 2.119) preschoolers had decreased and were statistically insignificant (t [6056] = 1.348, p = .128).[49] By 2013, average BLLs among white (M = 2.40, STD = 1.25) and Black (M = 2.51, STD = 1.605) preschool children were slowly increasing. But the gaps in their average levels of lead exposure remained statistically insignificant (t [1173] = 1.29, p = .198).

After the water source shift, though, this trend changed. During this time, the gap in the average level of exposure between Black children (M = 3.11, SD = 1.64) and white children (M = 2.80, SD = 2.32) expanded and became statistically significant (t [1284] = 3.09, p = .002). The percentage of children of both race groups with elevated BLLs increased between the pre-switch and the post-switch period. But the percentage of white children with lead poisoning increased from 3.3 percent pre-switch to just over 5 percent during the

Table 5.2. Children's mean blood lead levels by census tract in the City of Flint after the water source switch, 2013–2015

Census Tract	No. of Children	Mean	Standard Deviation
1	51	2.24	1.11
2	42	2.55	1.42
3	61	2.70	1.91
4	34	2.88	1.74
5	57	2.33	1.34
6	63	2.37	1.08
7	63	2.32	1.24
8	23	3.61	6.15
9	78	2.37	1.14
10	50	3.94	4.24
11	63	2.32	1.87
12	89	2.55	1.33
13	67	2.36	1.45
14	25	2.72	1.72
15	39	4.03	4.69
16	127	2.78	2.57
17	53	2.38	1.39
18	97	2.37	1.25
19	82	2.26	1.03
20	17	2.24	1.35
22	58	3.31	2.60
23	56	3.27	2.84
24	68	2.34	1.29
26	61	2.56	1.35
27	98	2.89	5.14
28	56	2.45	1.72
29	18	4.44	8.44
30	87	2.26	1.14
31	35	2.37	1.61
32	61	2.18	1.30

Table 5.2. (*continued*)

Census Tract	No. of Children	Mean	Standard Deviation
33	54	2.26	0.99
34	151	2.21	0.96
35	76	2.51	1.37
36	144	2.42	1.17
37	83	2.93	2.04
38	31	2.97	2.50
40	127	2.60	1.71
10304	103	2.65	1.41
10501	56	2.34	0.96
10502	24	2.67	1.83
10503	10	2.50	0.71
10504	12	2.25	0.97
10810	48	2.31	1.10
10811	63	1.89	0.95
10812	108	2.36	1.15
10813	96	2.14	0.96
10910	57	2.25	1.17
10911	97	2.35	0.84
10912	26	2.65	1.55
11010	100	2.23	1.02
11212	58	2.33	1.02
11213	11	2.36	1.36
All	3,852	2.49	1.966

Source: author's calculations of MDHHS data.

post-switch period in 2015, while the percentage of Black children with lead poisoning was nearly 4 percent (3.8) in the pre-switch period and peaked to just over 6 percent (6.4) in the post-switch period. What this means is that a statistically significant racial difference in BLLs existed within the state's data after the water source switch—data that a local woman, April Cook-Hawkins, then case manager in the Genesee County Health Department, claimed in Michael

Moore's documentary *Fahrenheit 11/9* was stripped of its most extreme cases. She maintained that local health department employees were directed to alter the blood lead test results of the kids in Flint to hide evidence of this silent but potentially deadly public health crisis.[50]

Even if these claims are true—that is, the highest test results were changed or deleted—the state administrators who monitor the public health data had known enough to develop a profile or mapping of lead-exposed children in Flint at the smallest geographical area, the census block. And it turns out that the city also could have taken into account the usual predictors of lead exposure to establish clear protocols for finding lead-poisoned children in the area.

Dangers Associated with Aging and Blighted Neighborhoods

The City of Flint had access to technical guidance from a host of departments and agencies, both internally and externally, that were aware of the current research concerning associations between neighborhood-level factors and lead exposure within high-risk communities. High-risk communities, scholars contend, are more likely to contain residents who live in concentrated poverty and within substandard housing that is surrounded by blight.[51] Researchers have identified that, in addition to residing near industrial sites that expose them to lead,[52] residents in high-risk communities face a greater risk of lead poisoning due to the presence of lead pipes and aging homes.

Effects of Lead Service Pipes

Well before lead gained widespread acceptance among cities and plumbing associations, the public health community had been aware that water from lead service pipes could be dangerous to vulnerable populations.[53] A Massachusetts State Board of Health report in 1898 indicated that the continued use of water flowing through lead service pipes was harmful because the water, which contained small doses of lead, "caused serious injury to health." This nineteenth-century report also illustrated that changing the source of the water seriously impacted the quality of water flowing through lead pipes, making the use of the water "dangerous to health where no danger existed before."[54] Concerns regarding the health effects of consuming water from lead pipes continued to permeate the public's consciousness well into the twentieth century, especially as researchers and public health professionals increasingly connected lead poisoning with lead water pipes.[55]

Despite the evidence illustrating the health effects of lead water pipes and products containing the metal, including lead-based paint, lead industry stakeholders have persistently denied these outcomes and criticized medical research. For example, instead of responding proactively to public concerns about lead, industry stakeholders opted to avoid disclosing the health risks associated with lead or discouraging use of the product. On the contrary, the lead industry aggressively promoted the use of lead paint for interior use in residential homes through Word War II, oftentimes targeting children in these public relations campaigns. Accordingly, as concern about lead-based paint re-emerged across the country in the early 1950s, the lead industry continued to maintain its innocence, claiming that the lead problem was caused by parental neglect. According to the Lead Industries Association, poor people residing in "slum dwellings" who improperly supervised their children were responsible for the lead poisoning problem.[56] In defense of lead-based paint, the association went on to claim that "until we can find means to (a) get rid of our slums and (b) educate the relatively ineducable parent, the problem will continue to plague us."[57]

Meanwhile, the lead industry has continued to utilize the tactic of misdirection to undermine calls to remove lead pipes. Despite the water system industry's awareness of lead water pipes in America's cities and towns, this industry, represented by groups including the Association of Metropolitan Water Agencies, the American Water Works Association, the National Association of Water Companies, and the National Rural Water Association, has consistently contended that lead poisoning derives from individual home plumbing as opposed to the water system's lead pipes and lead parts. While consistently sidestepping the topic of lead service line removal, the lead industry has adopted the strategy of redirecting attention to aging water pipes made out of iron and advocating for an increased commitment at state and federal levels to lead screening, product labeling, and public education.[58]

Effects of Living in Old Homes

Age of housing stock is another commonly recognized risk factor in state-level efforts to identify populations with a high risk for lead exposure.[59] Decades of research associate lead exposure in high-risk communities to residential status—in particular, whether a child lives in a house built before 1978, when there were minimal controls on the amount of lead in consumer products like paint, plumbing materials, and toys. It has been discovered that Black children are more likely than white and Latinx children to live in old homes with peeling paint and dust particles on windowsills,[60] elements that have been known to increase their risk of lead exposure.[61] Previous research also dictates that homes built before the 1950s are highly correlated with having lead service lines[62] and lead exposure from peeling lead-based paint and dust particles as well as with generally poor conditions.[63]

As noted in previous research, such as that of Richard Sadler and colleagues,[64] it turns out that the age of the homes that children resided in was associated with elevated BLLs in Flint. Consistent with such research, high lead levels are associated with living in homes built before 1950 in Flint.

Connections between Brownfields, Water Main Breaks, and Lead Exposure

My analysis of the private health data also suggests that living near hazardous exposures in this community, including brownfield sites[65] and water main breaks (which are forms of neighborhood blight and manifestations of uneven development),[66] matters.

Brownfield Sites

Children are vulnerable to brownfield sites because they are contaminated properties whose danger is oftentimes hidden from plain view. In most communities, brownfields are represented by abandoned and underutilized factories, junkyards, gas stations with leaking underground gas tanks, and other deserted industrial/commercial facilities that contribute to urban blight. These are the spaces where industry and small businesses left their mark by polluting the soil and water with arsenic and other chemicals. Due to the toxins found in the groundwater and soil, these contaminated properties are harmful to public health and have been connected to cancer, respiratory problems, and low birth weights.

Brownfield sites are everywhere in the United States. Authorities are currently aware of over 450,000 such sites nationwide.[67] Declines in the manufacturing industry took away family-sustaining jobs and left communities to deal with histories of uneven development. As a result, brownfield sites have become most prevalent in poor and discarded cities like Flint, where they are less likely to receive required remediation and redevelopment.[68]

Water Main Breaks

A water main is an underground pipe that carries water to customer service pipes.[69] Often, especially in cold Michigan winters, holes or cracks develop in the water main, then expand to the surface, creating extensive water flow at or near the site of the break until it is repaired.[70] Water main breaks are typically caused by the external corrosion of pipes during cold seasons when frost penetrates three to five feet into the ground.[71] They can cause extensive property damage and usually contaminate the water supply, even when the water appears to be clear.[72] Water breaks have been known to cause various ailments, such as gastrointestinal disorders, disease outbreaks, and illness associated with exposure to unsafe toxins and chemicals, including lead, and dangerous microbial contaminants, like E. coli.[73]

After a water main break is repaired, water may still be contaminated, especially in poor cities that lack funding to monitor the problem until the water is determined safe. Protecting residents from water main failures and minimizing the effects of water main breaks are primary challenges facing water utilities across the United States. In cities with severely deteriorated water distribution systems and a lack of funds to rehabilitate the infrastructure, such as Flint, water main breaks are rarely repaired.[74]

In recent years water main breaks have been reported frequently in Flint, with 378 in 2014 and 357 in 2015.[75] Visible standing water from breaks has been known to shut down parts of the city.[76] And, when water breaks cause water systems to lose pressure, they have also exposed residents to dangerous bacteria and prompted boil water advisories.[77]

Flint's consistent issues with water main breaks have a lot to do with the water system's maintenance history. Documents attained from the Michigan DEQ show that Flint had needed to improve its water transmission/distribution system since at least 2009. In an April 2013 water system distribution survey, the DEQ identified deficiencies in Flint's distribution system, especially on the city's west side, where a high concentration of elevated lead was discovered in September 2015.[78] In this 2013 water survey, the DEQ noted that main breaks occurred in many areas of the city; however, the most frequent occurred in areas where the distribution system had been identified as needing improvement: "Our main concern with Flint's water system continues to be the condition of the piping. Although the city has replaced approximately 12 miles of water main over the past ten years, much of the remaining piping is over 60 years old and in need of replacement. Also, this rate of water main replacement (1.2 miles per year-average) results in a total replacement rate of over 400 years. The typical water distribution replacement rate needs to be 100 years or less.[79]

Independent t-tests were helpful in determining whether living near brownfield sites and water main breaks matters when exploring lead-exposure differences in the Flint community. I learned from these t-tests that there is a statistically significant difference between mean BLLs among children living closer to water main breaks (M = 2.44, STD = 2.31) and brownfield sites (M = 2.48, STD = 2.81) compared with children who reside farther away from water main breaks (M = 2.27, STD = 1.75) and brownfield sites (M = 2.27, STD = 1.59). Overall, these results suggest that children with higher BLLs tend to live closer to these neighborhood hazards than do children without lead poisoning.

Map 5.1 shows the relationship between elevated BLLs and water main breaks. We can see that there is a significant association between lead poisoning and proximity to water main breaks. This map captures how the effect of living near water main breaks was more substantial for children within 500

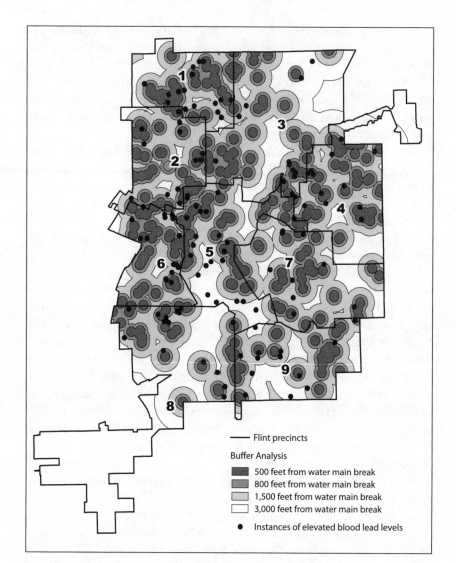

Map 5.1. Relationship between water main breaks and childhood lead poisoning

feet and that this effect declined out to 3,000 feet. This spatial pattern suggests that lead exposure rates are highest among children residing closest to water main breaks.

There is also a need to examine the relationship between brownfield sites, water main breaks, and lead contamination, given that pipe systems are not built to resist contaminated soils.[80] Contaminated soil has been known to seep into pipes (for example, cast iron, steel, and glass-fiber reinforced plastic systems), which are joined by natural or synthetic rubber gaskets.[81] This means that, in the long term, cast iron and steel pipes rust or crack, providing additional opportunities for contaminated soil to compromise public water systems.[82]

Other Factors Affecting Flint's Capacity to Protect Children with Elevated Blood Lead Levels: Follow-Up Testing

Since lead can ruin a child's health, affect development, and, at extremely high levels, even cause death,[83] physicians who find evidence of elevated BLLs are compelled to act, using available methods to intervene regarding both the sources and the effects of lead contamination. For this reason, follow-up procedures are essential in efforts to address abnormal BLLs in children, even in those who do not require intensive medical treatment. Typical best practices for follow-up care involve consistent and repeated BLL testing, risk assessment, case management, and lead exposure control.[84]

The Centers for Disease Control and Prevention recommendations for follow-up blood testing are based on the child's results in his or her first BLL test. That is, the degree of retesting varies according to the intensity of BLL elevation in the first test. For initial tests of 5–9 μg/dL, the CDC recommends one follow-up test every three months. As concentrations increase, the follow-up testing should be done sooner and more frequently. Children with BLLs greater than 45 μg/dL must be tested again as soon as possible. Once the child's levels have begun to decline, the CDC recommends follow-up testing at longer intervals and suggests that children with BLLs of 5–9 μg/dL continue follow-up testing every six to nine months.[85] This standard is half the time that health regulators gave physicians more than a decade ago. BLL tests were considered follow-up tests if they occurred before six months or 180 days lapsed.[86]

Despite these recommendations, each state has the autonomy to develop its own requirements. Consequently, there are huge disparities across states' policies. Five states lack any policies addressing the CDC's recommended BLL screening and follow-up testing at all: Arkansas, Montana, Wyoming, North Dakota, and South Dakota. These same states' websites lack information on childhood lead testing. Ten states plus Washington, DC, require universal childhood BLL testing and boast policies extending beyond the CDC's recommendations.[87] Massachusetts, which has some of the strictest protocols, requires health providers to administer follow-up tests for children with BLLs of 5–9 μg/dL within two months (rather than the CDC's one to three months or six to nine months if the child's BLL is declining).[88] According to the Michigan

Medicaid Provider Manual, BLL test results of less than 5 µg/dL require no additional action, but test results greater than or equal to 5 µg/dL necessitate a follow-up or confirmatory venous sample within three months, or, if the result is over 15 µg/dL, within one month.[89]

Private health data show that children's prolonged exposure to lead, even during a public health crisis involving the threat of lead-contaminated water, can be attributed to gaps in access to retesting. Efforts to verify that the results are accurate offer an opportunity to help parents avoid lead exposure risks in their environment.

Table 5.3 presents the demographic characteristics of follow-up testing patterns among preschool-age children living in Flint. It accounts for retesting that occurred within three and six months after the initial screening. Overall, as the table shows, there was a great deal of variation in the rates of follow-up testing in Flint during this time. Some children with lead poisoning received follow-up tests before three months lapsed (15 percent); however, the vast majority of children in Flint did not receive follow-up testing within three to six months.

Although public health laws require medical providers to target poor communities where lead poisoning is more prominent, none of the demographic characteristics that have been known to predict lead exposure—such as race, income, residing in homes built before 1950, and Medicaid insurance status—are helpful in predicting the children who received follow-up testing as specified by state and federal law.

For instance, although children with Medicaid make up over 96 percent of the population whose lead levels were retested, only 15 percent of these Medicaid-enrolled children received follow-up testing. Children with lead poisoning, as determined by their initial test, were most likely to receive a follow-up test when their BLLs were extremely elevated. As noted in table 5.3, children with BLLs at 10 µg/dL and over were more than four times more likely than children with BLLs between 5 and 9 µg/dL to receive a follow-up test during the Flint water crisis.

Previous research suggests that numerous hurdles exist to ensuring that children with lead poisoning receive follow-up testing as recommended by the CDC. For instance, some children fail to receive follow-up treatment because of the limited scope of medical services available where primary and preventive medical services are provided. Even when children with elevated BLLs return to health care facilities for other treatments, medical providers frequently fail to administer the appropriate follow-up tests.[90]

Children also reportedly miss opportunities for follow-up care for reasons that are beyond the control of health care providers. This is because some children, especially those from families that are poor and contending with housing

Table 5.3. Characteristics of Flint's lead-poisoned preschool-age children who received follow-up testing during the water crisis, 2013–2015

Characteristics		Distribution No. (%)	Percentage Retested within 6 Months	Percentage Retested within 3 Months	AOR (95% CI for AOR) of 3-Month Retest
Overall			21%	15%	—
Race	White	40 (42.1)	85	72	—
	Black	55 (57.9)	15	28	2.24 (.59, 8.56)
Moved	No	143 (86.1)	65	84	—
	Yes	23 (12.9)	35	16	.61 (.099, 3.79)
Median household income at parcel level under the FPL	No	118 (71.1)	79	68	—
	Yes	48 (28.9)	21	32	1.21 (.37, 3.97)
Blood lead level	5-9 µg/dL	131 (78.9)	56	56	—
	10 µg/dL	35 (21.1)	44	44	4.29 (1.38, 13.32)**
House built before 1950	No	32 (23.9)	19	26	.828 (.23, 2.97)
	Yes	102 (76.1)	81	73	
Medicaid recipient	No	132 (79.52)	3	4	
	Yes	43 (25.9)	97	96	1.18 (.11, 12.59)

Source: author's calculations of MDHHS data.

Notes: Prediction model also includes sample type, sex, age, age squared, and Latinx ethnicity. Percentages are rounded up. AOR = adjusted odds ratio of having received a retest within 3 months. CI = confidence interval. FPL = Federal Poverty Line

** $p < .05$

insecurity, are less likely to return for medical services after abnormal BLLs are detected. Missed opportunities for follow-up testing also emerge, in some cases, when parents, for reasons beyond socioeconomic challenges, neglect to comply with the physician's order for follow-up BLL tests.

A California assessment found that low follow-up testing rates were primarily associated with resource disparities specific to community settings. In this study, children were less likely to receive lead testing when they were referred out for screening. As a result, patients of clinics that did not have a pediatric phlebotomist on staff to draw the blood were far less likely to receive follow-up treatment when elevated BLLs were detected. This issue of referring children out for testing was compounded by parents' work schedules, lack of transportation to the testing facility, and the parents' minimal understanding of the importance of BLL testing to their child's well-being.[91]

While this analysis of follow-up testing patterns is not an attempt to determine the cause of low and disparate rates of follow-up care, it confirms that follow-up testing was underutilized during the Flint water crisis. It also raises several policy and program issues that should be considered as health officials in Flint continue to fine-tune their lead prevention and treatment efforts.

Trends in Lead Poisoning Surveillance Data Ignored

Evidence from health departments and providers in Flint pulls no punches: noteworthy differences in lead exposure existed in this community. BLLs among children varied by race and class. Some census tracts in the community had BLLs that were, on average, dangerously close to the State of Michigan's current standard for lead poisoning: 4.5 μg/dL.[92] However, despite this reality, the state said nothing and opted to obstruct justice by withholding knowledge of trends and associations between lead exposure and known sources of water contamination—knowledge that it had sole access to as gatekeeper of public health and environmental hazard data. The state could have been more forthcoming and helpful to Flint. But it wasn't.

Accordingly, it may be more politically viable to set aside the history of racial prejudice as well as the consequences of uneven investment and segregation in Flint. But, that's a dangerous choice. The roots and the fruit of neglect are connected. Failing to acknowledge the roots and the fruit of this level of depravity undermines any attempts to attain justice for Flint residents. This stance also pushes us further away from understanding the consequences attached to decades of neglect and disinvestment in predominantly poor and Black spaces.

Grasping this aspect of Flint's water quality issues does not take anything away from the victims, who certainly include all races, income levels, and age groups. Many white residents became sick from the water. Some lost their homes because they couldn't pay for water that they couldn't even drink. They were also being forced to pay high rates for tainted water that they couldn't bathe in or feed to their plants and pets—just like the rest of Flint's residents, especially in areas with a high concentration of water quality issues.

The truth is that the flow of toxic water in Flint was not choreographed to impact Black residents in particular. But it is important to note that this population disproportionately resided in areas overburdened by neighborhood disorder that affects water quality.

The cavalry came to Flint by the end of 2015. Journalists, academics, politicians, and environmentalists descended upon the city, and resolution seemed in sight. But health administrators and water regulators continued to sit on important information and prevented the creation of more timely and targeted remediation plans. And the politics of neglect persisted without interruption in Flint. So more months flew by before Flint would be forced to deal with the impact of the lead crisis.

It took even longer for the cavalry to show up in the city's predominantly poor Black areas. And before any of that happened, alliances were questioned and tensions surfaced among water activists vying for control of what had become a significant environmental justice story in this country.

These tensions began to add to other pressures, including beefs among city administrators and the constant bickering and infighting that played out among Flint's city council members at most city council meetings. This chaos also threatened to undermine the fortitude of citizen-led efforts to repair Flint's water infrastructure.

6

The Blame Game

A Legal Circus and Public Finger Pointing

"I'm sorry and I'll fix it. . . . I let you down Flint," was what then governor Rick Snyder said in his State of the State address on January 19, 2016. He went on to pledge that, moving forward, no one in Flint would go without access to clean water.[1]

While Snyder was delivering his speech in the Capitol Building, hundreds of protesters were rallying on its east steps on a Tuesday evening after a day spent at Lansing City Hall. Bundled in winter coats, hats, and gloves and enduring freezing temperatures, they braved the elements to oppose Snyder's decision making in Flint. Holding megaphones and ringing cowbells, some of the protesters chanted "Jail to the Chief," while others joined in with "Hey, hey-ho, Snyder's got to go." There were signs reading "I Stand for FLINT," "FIRE SNYDER," and "Snyder: What did you know? When did you know it?" These concerned citizens and elected local leaders represented various segments of the community who were fed up with the state's inaction and were calling for the governor's resignation.

It didn't take long to discover that Snyder's apology did not quiet the backlash that had erupted after the State of Michigan acknowledged not only the dire reality of Flint's water issues but also the ways state officials had mishandled the situation. Citizens had only recently discovered that Snyder-appointed emergency managers could have avoided destroying their water system if they had opted to spend only $100 a day for corrosion protection. State-appointed leaders had also failed to ensure that the city's water treatment plant had the capacity to clean the city's water supply. Learning that the state wouldn't carry out even these basic and vital steps that would protect their health and well-being and instead had switched to a water source that had long been treated as the city's local dump—namely, the Flint River—was infuriating. For citizens, the deliberate policy and regulatory mismanagement that had led to the contamination of an entire city's drinking water supply was completely unacceptable. Public officials, from a citizen's perspective, had betrayed their city and deserved to be in jail for their crimes. Water rights protesters wanted justice.

The first step on the path of justice involved, at least from the state and federal perspective, finding out what had happened in Flint. This would lead to the person or organization that had caused this public health disaster. A US congressional hearing in mid-February 2016 aimed to find such answers. Individuals who had helped expose the city's lead-contamination issues testified about their experiences, the health effects they had suffered, and the frustrations of trying to work with state and local water regulators. Lee-Anne Walters, for example, testified that it took the city too long to take her concerns seriously. Regulators behaved as though she was wasting their time and making unreasonable claims—until her doctor confirmed that consuming city water had damaged her health:

> As we were showing them our water, I was told that I was a liar, and I was stupid by showing these bottles of water. Saying it did not come from my home and that I would not get anybody to believe that that was my water. City of Flint started coming in and testing in my home in February of 2015 after the doctor's notes. While the city was coming in and testing my home, they were seeing on a weekly basis that the brown water was something that had become consistent in my home since January.[2]

Even after Flint residents provided proof that their concerns were valid, any resolution to their problems continued to be slow to develop. Walters had been so upset, she told the congressional panel, that she called a civil engineer at Virginia Tech, Marc Edwards, in tears. She sobbed that officials seemed comfortable letting "an entire city be poisoned." She recalled that, because of concerns about pushback from officials, residents who were having their water tested implemented ways to protect the results from being discredited, including following a suggestion from her husband to have only homeowners open the testing kits upon receipt from Virginia Tech and then return them directly for independent testing.

Nonetheless, the state's Department of Environmental Quality questioned the community's water results and testing procedures. Walters was incensed that officials were openly trying to discredit rather than help Flint's families. "When does this stop?" she asked. "When do the citizens get what they need? When are we gonna be heard in the manner of which we deserve? The programs that are out there right now that assist people. My family was not offered any services. For anything. I have a child who is lead poisoned; I have children that were exposed.... We all have health issues at this point."

Researchers also spoke at the hearings. They, too, had helped expose the water crisis. Dr. Mona Hanna-Attisha, whose work was instrumental in efforts to correct water quality, testified that she had been incredibly disappointed that the city and the county both seemed disinterested in her findings. Rather than alert the necessary parties, they passed the buck, one after the other, claiming the issue was some other department's responsibility:

> I was told that the County Health Department has no jurisdiction over water. That's under the control of the Public Works Department. I kind of scratched my head. I'm like, "Well, this is a public health issue, how can you not have control over the possibility of lead in water?" I was just told it's not under our jurisdiction. That day, this was August 27th, I emailed the leadership of the Health Department, including the environmental health supervisors, reiterating my concern about lead in the water. A few days later, [I] was told again [that] this is a concern, but this is not our jurisdiction. That kind of started my crusade to get data about blood/lead levels.

For their part, Genesee County health officials testified that their hands had been tied. Mark Valacak, who had retired from his position as a Genesee County health officer in October 2017,[3] defended his response to Hanna-Attisha's pleas for assistance, testifying that in Michigan "we no longer have a state department of public health.... The programs that were once in the department have been spread across four different departments of state government. So the public health system is complex and involves multiple agencies on the local, state, and national level. It is difficult for someone who works in public health to understand all the pieces and explain who does what. The public is even more challenged in trying to understand the system." In addition to the bureaucratic mess within the county's public health departments and agencies, state officials—including former head of EPA Region 5 Susan Hedman, emergency manager Darnell Earley, and Governor Snyder—assigned blame for the Flint water crisis to, well, *anyone but themselves.*

Meanwhile, former Flint mayor Dayne Walling's account contradicted their hollow and noncommittal stances. He argued that the emergency manager system had helped create the crisis. "From my experience as mayor until November of 2015, Michigan's financial manager system focused too much on cutting costs without adequate safeguards and transparency. Second, the city did not have the experience and capacity in place to manage a new water source. The Flint Water Taskforce report was very clear on that. The regulatory agencies also provided false assurances to us about the safety of the water and with-held information about the risks. Last, local concerns were discounted, and our agencies did not act with urgency."

Walling adopted the same rhetoric of remorse and concern displayed by other speakers, pleading for increased transparency in the water management system.

> A big part of what I see in retrospect is that we didn't have that inherent kind of check and balance with the public that you have in normal, home-rule situations. As I look back and try to figure out how I got the informa-tion I got at different points under this Emergency Manager about this problem, I had every expectation when I was sitting down with the DPW [Department of Public Works] director after the order was in place for that person to report to me for day-to-day responsibilities. I had every ex-pectation that I was getting the same kind of critical information to make decisions that I would get if the Chief of Police were providing me with a report. If you have a spike in burglaries, then you bring that up the chain-of-command. There was, in this case, just a lot of information that was being parceled out and withheld. That was compounded by the fact that there wasn't the regular public disclosure and public discussions that take place during your normal five-hour City of Flint City Council meeting.

The congressional hearings were an intriguing public exchange. The appropri-ately angered congressional leaders shook their heads at officials who had let the crisis happen. They expressed grave concerns about the future of Flint. Yet direct assistance to the city's residents—in the form of restitution for this pub-lic health meltdown or remediation to stop the damage from continuing—was not forthcoming.[4] The hearings documented how the crisis could have been prevented with adequately staffed health and water regulation programs with clear lines of authority and oversight and open access to local data, but they did not do a whole lot about the mess moving forward.

The hearings did not alleviate exorbitant water bills or expedite efforts to fix the water supply. Instead, they turned out to be a well-orchestrated display of righteous indignation for members of Congress—who had the opportunity to express their disgust with the treatment of the residents of Flint. They also put on display the blame-avoidance carried out by local, state, and federal of-ficials who did nothing but implicate one another in this public health disaster. But in the end, while Americans had an opportunity to watch state agencies deflect blame for events they held at least some responsibility for, the hours of testimony—as well as the attempts to compel Governor Snyder and emergency manager Darnell Earley to testify—were a wasted effort.

What these hearings accomplished for Flint was what congressional hearings accomplished for Washington, DC, when it was dealing with its lead crisis in the late 1990s—absolutely nothing. Environmental justice advocate Erin Brockovich said it best when she described the Flint water crisis hearings as "frustrating" and a "waste of everybody's time."[5]

Despite Flint's persistent presence in the national spotlight—due to the well-publicized wrongdoings of government regulators and public administrators—in this chapter I will document why justice sought by water activists eluded them, even after President Obama's visit to the city on May 4, 2016. This was in part due to the constant conflict among the people who had helped bring attention to Flint, which upended the collaborative effort between prominent researchers and citizens to fix the city's drinking water. Meanwhile, efforts to hold public administrators and water regulators legally responsible—under both criminal and civil law—can best be described as stunted and inconspicuous.

President Obama Rides into Town

Months after Governor Snyder issued his apology and the congressional hearings had come and gone, outrage among residents and their supporters continued to fester. How had the State of Michigan managed Flint's water system? The information that continued to emerge didn't make the state look good. For instance, emails released after Snyder's apology revealed that—among other indiscretions—state officials had made provisions to protect the health of state employees working in Flint from contaminated water before bothering to warn the citizens using that water system to stay away from it.[6]

Just as Flint residents were getting accustomed to receiving empty promises from water regulators and government leaders regarding their plight, the White House announced in the last week in April 2016 that President Barack Obama was coming to Flint! Citizens were overjoyed that the first Black president—who had won America's heart with his campaign slogan "Yes We Can"—was on his way to their city. The people of Flint already had science on their side. Now the president of the United States was headed to town to bear witness to their truth.

Certainly, some residents thought, President Obama understood how urgent it was to address the city's water problems. They believed he would exercise his authority to compel the State of Michigan to right wrongs. Flint native and US representative Dale Kildee, who represented a suburb of Flint Township that had been accused of "resource hoarding,"[7] expressed gratitude for the president's visit: "I thank President Obama for keeping the focus on Flint families affected by the city's water crisis and I look forward to welcoming the President to my hometown."[8]

Following the arrival of Air Force One in Flint, Obama was met by Governor Snyder. Many residents waited for the president outside of Flint Northwestern High School, where he was scheduled to speak with the community, reasserting their demands with bullhorns and picket signs. Others were "humbled and excited" by the president's trip while seeming to have no explicit expectations.[9]

When Obama showed up, he didn't disappoint—initially. He spoke with residents and met with community leaders, all of whom expressed their concerns and hopes for federal intervention. He opened his speech by telling them about a letter he received from eight-year-old Amariyanna "Mari" Copeny, a Black girl from Flint who had written him in March 2016, which had inspired his April 25 response—and his current trip to the city. He said to Flint residents,

> Like a lot of you, Mari has been worried about what happened here in Flint. She's worried about what it means for children like her. She's worried about the future of this city and this community. . . . Now, I would have been happy to see Mari in Washington. But when something like this happens, a young girl shouldn't have to go to Washington to be heard. I thought her President should come to Flint to meet with her. And that's why I'm here—to tell you directly that I see you and I hear you, and I want to hear directly from you about how this public health crisis has disrupted your lives, how it's made you angry, how it's made you worried.

He went on to tell those in the audience to keep their heads up because he had "[their] backs" and that "we're paying attention." Obama concluded by trying to rally Flint residents, predicting that the city would "bounce back not just where it was, but stronger than ever."[10]

Some vocal Flint residents and supporters, like filmmaker and Flint native Michael Moore, were not happy with Obama's speech. Moore told a CNN reporter that he was not especially impressed by the "anecdotal stories" and wishful thinking that he heard from the president concerning the city's capacity to rebound if there was not a commitment of federal support to make that happen.[11]

There were also residents and supporters who claimed to be disgusted by what some have referred to as Obama's preliminary "stunt," which include his public consumption of filtered City of Flint water.[12] This display was especially problematic because, for months before his visit, residents had been claiming that the filters provided by the state did not adequately clean their water.[13]

While delivering his speech, President Obama set up his public display of drinking Flint water by letting the crowd know that he was thirsty, a move that was reminiscent of Obama's decision to take a plunge in the Gulf of Mexico with his daughter Sasha after the area was declared "safe . . . and open for business" following the BP oil spill in 2010.[14] When the crowd erupted with concern, he told them, "I'm all right. I'm gonna get a glass of water right here. Let's make sure we find one. . . . Settle down everybody." But no one brought the president water. Maybe the person responsible thought that he was joking. This prompted President Obama to ask again, this time coughing. "I'm still waiting for my water. Somebody obviously didn't hear me. Usually I get my water pretty quick."[15]

Moore, who mentioned that he was not overly excited to be critical of the president, said that he should have just stayed in Washington, because his visit would not help Flint. "He walked out there blue-collar [with his] jacket off and sleeves rolled up," Moore said. "[He] looked good and sounded good. [But] tomorrow morning, the children of Flint [will still be] drinking the same poisoned water."[16]

But there were also plenty of Flint residents who were not critical of Obama and his visit. And they weren't upset that it did not drastically change circumstances regarding the city's water problems.

"We wasn't into that whole political stand about what Obama was doing. No, it's about what they were doing *before* Obama even came to Flint," a resident told me, adding, "Flint's been fucked up for a long time. Why do you think they didn't go to these other places like Southfield and pull their switches? But they'll make us drink out of the Flint River."[17]

She told me that she was upset about the water like everyone else, having learned that there were problems with it two weeks after getting her teeth pulled by a dentist. As she recalled, when authorities made their announcement, "the water had been fucked up over a year. We would turn on the faucet and say, 'There's something wrong with this. It shouldn't be bubbling.' But, it seemed like you could put a light to it and the shit would catch on fire."

There were people who were "pissed off" with Obama, she knew, and she recalled an interaction with another Flint resident on a train headed south. The person said that Obama had lied to the city.

"But, let me tell you something. If he lied, it wasn't intentional," she stressed. "You know how the man is. Obama ain't never even had a speeding ticket. What the fuck are you talking about? *He* lied? It's too many people involved for all that finger pointing at Obama. So, other citizens of Flint might have different reactions. But *my* people, *we* ain't pissed with Obama because of what *the state* did to fuck up our water."

I wonder if the folks who were bothered by President Obama's visit would feel differently if they knew that young Mari Copeny wasn't the first resident to contact him about the water. A year before his visit, I remember listening to Melissa Mays recall how fed up she was with the response by water regulators and public administrators. Their disinterested responses, she had told me, infuriated residents and propelled them to seek people in authority. Following this conversation, I made attempts to gain access to these complaints from the DEQ and the EPA.

As we have seen in previous chapters, residents had actually been contacting the White House for almost two years before Obama showed up. Freedom of Information Act documents released by the EPA reveal that in addition to reaching out to the local water department and the DEQ, Flint residents sent their complaints to the EPA and other federal agencies, in some cases directly to the White House, in hopes of motivating federal authorities to leap into action and get them the support they needed.

Residents started contacting Obama in 2014, sharing with him what was happening in Flint. One wrote to the White House in January 2015, "Please excuse my grammar. I am not very good with it. But, will you please look up Flint, Michigan and see what is going on? We got bad water. It got THMs in it. The water system had been a mess for months now. Please, please we need your help because the mayor and the local government people are saying they cannot afford to switch back to Detroit Lake Huron water. The kids get rashes from THMs, and some pets died from it. We need help, please."[18]

Residents who wrote the president wanted to know how this could happen to them. They also wanted to know how this situation could be resolved. One resident, who signed off as "Fed up in Flint," wrote on January 27, 2015,

> How long must the citizens of Flint go without water? . . . Who do we as citizens have to stand up for us when our local officials will not stand up and take action? The water has been cloudy, even yellowish in color, with an extreme bad odor for months, yet for weeks they told us the water was fine. Only to retract their statement and say that the water is not fine and that long use of the water will be harmful to a baby, pregnant persons, elderly and [the] sick! Nobody will take a stand, so I decided to start here. Hopefully someone will hear my cry, not only for myself and my family but for the other citizens of my city who seem to have no one that cares.

Residents continued to send emails and letters throughout 2015 and 2016, wanting Obama's help because they assumed they were being treated unfairly due to the reputation of the city and its people. One resident pleaded on February 10, 2015, "Please help us. Everyone turns a cold shoulder . . . [but] this is America, not a third world country."

It would be another year before the president directly engaged the community. Prior to Obama's visit, the White House forwarded complaints to EPA Region 5. When Tinka G. Hyde, the director of the Water Division in Region 5, sent form letters in return—oftentimes several months after Flint residents' initial complaint to the president—the correspondence did not include any direct responses about why the water came out of the tap brown, smelled like sewage, and appeared to be causing rashes and contributing to the premature death of pets. Instead, Hyde thanked concerned residents for their correspondence and assured them that, despite recent violations, Flint had "returned to compliance" and that "all other regulated contaminants [met] State and Federal guidelines."[19]

All Politics Is Local: Challenges to Flint's Recovery

To be sure, positive developments did emerge from the news and media coverage and also from the congressional hearings and President Obama's visit. In response to the residents' call for help—and supporting the validity of their concerns—deliveries of bottled water poured into the city.[20] The United Way and the American Red Cross, alongside faith congregations and community groups, traveled hours to bring supplies and support the people of Flint. Churches, schools, and public buildings took precautions by covering water fountains with plastic.

Differing Opinions and Withheld Funds

Councilman Eric Mays said that circumstances in Flint appeared to be headed in the right direction. But, Flint would need an infusion of funds to continue that progress. According to Mays, "Anytime you can get money in to fix certain parts of an infrastructure, that's a move in the right direction; when we got them to switch off the river water and go back to Detroit in the interim, that was a move in the right direction. And anytime you can continue to make upgrades to the water treatment plant, that's a move in the right direction. So, yes, there's been some moves, no matter how big or small, in the right direction."

By the time of my conversation with him in late May 2016, the City of Flint was beginning efforts to replace miles of pipeline. Meanwhile, the EPA was declaring that the most significant problem was being solved due to the proper use of water filters. Some people found comfort in these official declarations that proclaimed that peak exposure had passed. But others, like Mays, weren't so sure that the water supply was safe—even after it was coming from Detroit. Circumstances varied around Flint, and Mays mentioned that, as an elected official, he was not "recommending that [residents] drink the water at this point."

Circumstances and opinions about the issue were varied because residents were forced to believe two conflicting realities simultaneously. Either the water was "all in the clear," as reported by the EPA, or, as Councilman Mays put it, there was still a great deal of work to do to improve water quality in Flint.

"So, you change the lead service lines," Mays related. "[But] I don't know what people will continue to do unless each house is properly monitored and inspected."

Flint residents were still using bottled water rather than public tap water when Mays cautiously described the uneven success of lead remediation efforts, which stemmed from substantially varying circumstances across the community. "Some people," Mays recalled, "have lead levels that the filters can handle, and some have higher lead levels.... So, there are class action lawsuits that will carry on. Some people had problems with rashes on their skin, hair loss, and so forth. And I didn't say 'alleged' rashes and hair loss because I witnessed some of this. I have not had the issues of bathing and showering and dealing with the water; but, some have. Every household has had a different experience."

I initially caught up with Mays in Flint's city hall after he announced his bid for mayor. He explained that he was on a mission and didn't have much time to chat as we walked past the city clerk's office. But I followed him around anyway—a former Flint Truck and Bus factory worker—in May 2015 as he tried to appeal the city's finding that some of the signatures on his candidate petition were invalid. For hours, we sat with county clerk staff and sometimes one of his supporters while auditing or reviewing signatures. At that time, he was exercising his right to challenge, within seven days, the excluded names. Mays was determined to review all excluded signatures.

Due to the confusion over the deadline for petitions, there were only "three" mayoral hopefuls whose paperwork was submitted on time. In fact, only Councilman Wantwaz Davis—who had been elected to the city council in 2013 despite having served nineteen years in prison after pleading guilty to a 1991 second degree murder charge—and Mays made the deadline. Well, they were the only *human* candidates to make the deadline. Trial attorney Michael Ewing also got Giggles the Pig's petition in on time. Ewing claimed that his nine-month-old pet was a better candidate than either Mays or Davis, seeing as she had no criminal record and could meet deadlines.

Though the porcine contender garnered laughs, she ended her mayoral bid on June 9 with a web post explaining that her candidacy had been intended to avoid the "write-in election" that might have been triggered if no valid candidate petitions had been submitted prior to the state's deadline. Giggles no longer needed to run, since a subsequent law allowed candidates' names to be on the ballot despite having submitted their petition late.[21]

Mays told me that the "signature thing" was just a "smokescreen" designed to undermine his attempt to become mayor. And while he was eventually able to legitimize his August 2015 run, he didn't secure the job—Karen Weaver won, with Mays coming in third. He was, however, one of very few city council members to maintain his seat in the election. When we spoke again, toward the end of May 2016, Mays attributed his constituents' loyalty to his devotion to truth telling and advocating—loudly and boldly—for legislative transparency, for improved living conditions for his ward's people, and for other resource-starved spaces in Flint. Like Bill Hammond, Mays seemed to have a special love for the city, the kind that one pours on a truant child that still has a heart of gold. He remembered Flint during its more prosperous years, though he was less nostalgic than Hammond and more prone to delivering soliloquies about restoring Flint to its greatness.

Acknowledging all the support that came Flint's way around the time of the congressional hearings, Mays admitted that progress was moving slowly. Fixing the water crisis was a sluggish endeavor, slowed by various factors—including attempts to make the best decisions for Flint while the community was still distracted with their ongoing water situation:

> You got some people working together, individuals, organizations, and so forth. But, when you look at the fact that the pipe that connects Flint to Detroit water was sold to Genesee County and see the inner workings of the Karegnondi Water Authority, you'll see that Flint is vulnerable. If Flint misses a $7 million bond payment, the county can take over the assets of the Flint Water Treatment Plant. So, while it appears as though people are working together, things are really divisive. You got people who believe water crisis money should have come to the city then; they distribute it to do what they please. But, the state has been withholding the money. I see this as intentional neglect of resources coming into the city. Just that old stigma [designed to] keep the local government down.

Water Justice Advocates Can't Get Along

Councilman Mays also mentioned that problems were brewing among the water activists who helped bring attention to the crisis in Flint in the first place. A collective of out-of-town researchers had been crucial in focusing attention on Flint's water problems, yet residents began to voice concerns with the tactics, comments, and conclusions made by some of them, especially Dr. Marc Edwards. As Eric Mays put it, "Everybody played a role in the water crisis. Lee-Anne Walters played a role. Dr. Edwards, Mayor Weaver and Melissa Mays [played their roles]. There's a whole lot of individuals that were involved. Ain't no one individual Martin Luther King [continuing as he thought aloud about internal arguments among the water activists].... Some of them had got star-struck. The media has had people fighting to be the Martin Luther Kings and Martin Luther Queens."

By mid-2018, tensions had bubbled over. In July, Edwards filed a lawsuit against Melissa Mays, Paul Schwartz, Yanna Lambrinidou (Edwards's former research assistant), and other leaders of Campaign for Lead-Free Water. Their once-fruitful relationship had deteriorated over time, with these activists alleging that Edwards's decisions were undermining the community's calls for water justice. In response, Edwards claimed—while seeking $3 million in damages—that Mays, Lambrinidou, and Schwartz had defamed him, ruined his professional reputation, and made false claims and statements concerning his work in Flint.[22]

Although Edwards helped Flint residents get their water tested, personality differences gradually undermined their collective efforts to resolve water concerns. His approach, according to residents who had worked with his team, was paternalistic and egotistical. Residents asserted they had no voice in their "collaboration" and that Edwards depended on them to check in with him, rather than his being proactive. This top-down method of community engagement distanced Edwards from the people whom he had initially shown up to help. Residents like Claire McClinton mentioned to ethnographer Benjamin Pauli that his presence had become "demoraliz[ing]" and that his need to be the ultimate authority, even regarding subjects that were outside the scope of his expertise, was exhausting.[23]

As the rifts began to surface, Edwards seemed especially infuriated by the allegations of professional misconduct that were included in a May 2018 letter distributed to associations and individuals within the scientific and engineering community. These included prestigious grant funders and Edwards's employer, Virginia Tech. The letter, posted on Flintcomplaints.com, contained allegations by water activists that Edwards had acted unethically.

Despite having led the team of volunteers and citizen scientists who helped document and expose the water problems in 2015—and later exposing the corruption among state-level water officials through FOIA requests—Edwards's credibility, for the letter writers, was shot. By January 2019 their complaint boasted approximately ninety signatories. The text read, in part, "Residents of Flint request you tell us where we can file a formal complaint against the behavior, since January 2016, of Professor Marc Edwards of Virginia Tech. Many of [us] feel that Mr. Edwards' drama, changes in stance, and attacks on residents and researchers have ended up taking Flint residents' voice away and giving it to Mr. Edwards. This has allowed Mr. Edwards to make Flint's Water Crisis about himself and not the people." Casting Edwards as a "disruptive presence" who "distracts from the real suffering in Flint," activists claimed to be shocked that he had "gone as far as to declare that the Flint Water Crisis was over 2 years ago (in 2016)." They stated it was necessary that Edwards "leaves our town alone" due to the confusion and "rifts he has created between residents" so that they could "try to address the real problems plaguing [them]."[24]

Edwards's conduct is described as "glib, reckless, and egotistical," particularly his contention that Flint was no longer plagued by harmful levels of bacteria in its water and that residents might have caused some of their health issues (including shigella, or dysentery) by practicing poor hygiene. Both claims opposed research from Professor Shawn McElmurry and the Flint Area Community Health and Environment Partnership (housed at Wayne State University).

Countering in his Virginia lawsuit, Edwards claimed that he never misled the City of Flint or said that the water crisis was over. Instead, he argued that once his tests began to resemble estimates reported by the State of Michigan in September 2017, he was willing to vouch for filter use. Further, he claimed that he did not state that the water was safe to consume or bathe in without significant caveats. The defendants named in the lawsuit, Edwards added, "purposefully and repeatedly misquoted [him] and intentionally and falsely attributed damaging statements to him that he never made."[25]

Attorneys for both sides parried, but a Virginia judge dismissed the lawsuit in March 2019. At the time of that decision, Edwards seemed to vacillate between claims that he planned to appeal the judgment or to drop the complaint. "Maybe it is my fate, to be the last person, to adapt to a post-truth world," Edwards told Michigan Radio reporter Steve Carmody. "Maybe I should just embrace that and go on."[26] The Flint residents and supporters who signed the online letter of complaint would probably agree.

Trying to Find Someone to Hold Accountable

The office of the Michigan attorney general joined efforts to get justice for Flint residents when it filed criminal cases related to the water crisis. On June 14, 2017, Attorney General Bill Schuette announced that, in addition to charging Michigan Department of Health and Human Services officials (including Chief Medical Executive Eden Wells) with lying to a peace officer and obstruction of justice, his office was leveling felony charges against MDHHS director Nicolas Leonard Lyon, the DEQ's Office of Drinking Water and Municipal Assistance chief Liane Shekter-Smith, former Flint emergency manager Darnell Earley, former City of Flint Water Department manager Howard Croft, and former DEQ water supervisor Stephen Busch. Charges ranged from conspiracy to willful neglect of duty to involuntary manslaughter.[27]

Lyon received a great deal of attention in these filings. He was charged with one count of involuntary manslaughter in the December 13, 2015, death of Flint resident Robert Skidmore, with failing to protect City of Flint residents from an outbreak of Legionnaires' disease, and with grossly negligent conduct with regard to investigations of the outbreak. That second charge was official misconduct, described by the attorney general as efforts to "intentionally mislead and with[hold] information about the Legionnaires' disease outbreak [and] direct a health official to discontinue an analysis that would aid in determining the source of the Legionnaires' disease outbreak and save lives."[28]

Jeff Seipenko, the special agent assigned to the Office of Special Counsel responsible for the Flint water criminal cases, interviewed hundreds of witnesses, surveyed twenty depositions secured via investigative subpoena, and reviewed reams of emails and other correspondence. Seipenko reported that Lyon had first become aware of the Legionnaires' disease outbreak on January 28, 2015—but then waited a year to inform the public of the outbreak.

Perhaps more damning, he found that Lyon had been approached by MDHHS state epidemiologist Corinne Miller about an uptick in Legionnaires' cases in 2014. There had been no more than ten cases reported in Genesee County in 2013. But by October 2014, Miller reportedly informed Lyon that there were thirty cases—which she thought constituted an outbreak. When Miller met to discuss this matter with Lyon on January 28, 2015—armed with statistics illustrating a threefold increase in Legionnaires' between 2013 and 2014—she told Lyon that her department could not rule out the Flint River as the possible source of the outbreak. Miller again alerted Lyon, through his executive administrative assistant Nancy Grijalva, after a second wave of Legionnaires' disease hit Flint in the summer of 2015.[29]

According to the Office of Special Counsel's report, Lyon also allegedly directed his employees to undermine accounts of elevated blood lead levels in Flint a month before Dr. Mona Hanna-Attisha's September 2015 study on statistically significant blood lead level increases. According to Linda Dykema, director of the Division of Environmental Health at MDHHS, Lyon asked staff to put out the word that the elevated blood lead levels in Flint's children were just the result of seasonal fluctuations.[30]

Regulatory officials within Michigan's DEQ compounded the duplicity coming out of Lyon's office. Shekter-Smith had been head of its Office of Drinking Water and Municipal Assistance. At key moments during this public health crisis, Shekter-Smith and other state regulators appeared more concerned about the Flint River being recognized as the source of a Legionnaires' outbreak than about protecting Flint citizens from the actual public health threat.

According to Harvey Hollins III, director of Governor Rick Snyder's Office of Urban Affairs and Initiatives, the DEQ was *also* aware of the Legionnaires' outbreak months before city residents were informed.[31] Much of the DEQ's joint efforts with the MDHHS seemed to be focused on trying to force an area hospital, McLaren, to take responsibility for the Legionnaires' outbreak. Julie Borowski, the hospital's compliance director, even claimed that the hospital received a letter from Lyon in which it was ordered to correct the problems because McLaren—and not Flint's water system—"is a nuisance, is in an un- sanitary condition and is a possible source of illness."[32]

Before and after the water source switch, Shekter-Smith was playing an es- sential role in securing the Administrative Consent Order issued by the DEQ to the City of Flint on March 20, 2014, which required the city to use the treat- ment plant and "undertake the KWA [Karegnondi Water Authority] public in- volvement project or undertake other public improvement projects to continue the use of the Flint River." In late March 2013, just over a year before the water source shift, Shekter-Smith was copied on an email sent by the DEQ's Stephen Busch to DEQ director Dan Wyant detailing the health risks of using the Flint River as a drinking water source.

Michael Glasgow and others at the Flint Water Treatment Plant repeatedly stressed, in conversations with the DEQ, the importance of adding corrosion control treatment before the switch. According to the attorney general's war- rant for Shekter-Smith, DEQ officials "stated to Glasgow that the City of Flint did not need to add any corrosion control treatment to the water."[33] Shekter- Smith did not make efforts to connect known water quality issues to the citizen complaints that had emerged less than two weeks after the water switch,[34] nor did she advocate for corrective measures to resolve complaints. On January 7, 2019, Shekter-Smith pleaded no contest to a "disturbance of a lawful meeting" misdemeanor and agreed to help the prosecution in order to avoid the origi- nal felony charges. Special prosecutor Todd Flood praised her "candor and truthfulness."[35]

By February 2019, a total of fifteen state regulatory officials, health scientists, and City of Flint utility workers were charged for their roles in manufacturing the water crisis. Among them, seven have pleaded no contest to misdemean- ors and agreed to testify against other defendants in the criminal cases.[36] As part of a plea deal with the attorney general's office, the water treatment plant's Glasgow admitted in Genesee County District Court on March 22, 2018, that he sent falsified Lead and Copper Rule water testing reports to the DEQ in 2014 and 2015 in order to conceal elevated lead results.[37]

Michael Prysby, the DEQ engineer who approved the water switch despite the known risks, was initially charged with two counts of misconduct in office and one count of conspiracy to tamper with evidence. In his plea hearing on December 26, 2018, Prysby agreed to testify that, before the water source shift, the necessary tests were not run to ensure that the water treatment plant could adequately treat Flint River water.[38] In exchange for one year of probation and having the charges dismissed, Prysby indicated that he was prepared to testify that the decision making to utilize the Flint Water Treatment Plant before it was actually ready was supported and approved by the two state-appointed emergency managers sent to Flint.

Claiming that the special investigator, attorney Todd Flood (working under the umbrella of former Attorney General Schuette), botched the investigation with a "flawed foundation" and questionable prosecutorial tactics, Dana Nessel—who became Michigan attorney general at the beginning of 2019—decided in June of that year to drop all pending cases against state officials and launch a new expanded criminal investigation, one that would take into account new evidence, including "millions of documents and hundreds of new electronic devices."[39] However, at the time of writing this book, the State of Michigan had yet to file any new criminal charges against water regulators and health administrators who allegedly contributed to the water crisis. With the highest-ranking officials still awaiting trial, Nessel and Wayne County prosecutor Kym Worthy—who joined the team prosecuting the Flint water crisis criminal cases in early 2019—assured Flint residents that new criminal charges for misconduct in office by a public officer would be forthcoming before the statute of limitations ran out.[40]

To support this effort, Flint's representatives in the Michigan legislature made attempts to extend the statute of limitations—from six to ten years—with the introduction of Senate Bill 462 (sponsored by Senator Jim Ananich) and House Bill 4834 (sponsored by Representatives John Cherry, Sheldon Neeley, and Tim Sneller).[41] Some Michigan legislators, like Republican representative David LaGrand, have voiced their concerns with extending the statute of limitations: "We have a very strong impulse to find somebody to blame. . . . I'm deeply reluctant to assign criminal blame to people who screwed up—to people who acted without malice." Unfortunately, for Flint residents eager to see the people who manufactured this public health disaster brought to justice, both bills languish in committee and will likely see no action.[42]

In the end, despite Flint residents' consistent pleas for justice, Michigan's first criminal investigation into the water crisis was a debacle, resulting in seven pleas of no contest to misdemeanors and the distant hope that charges will eventually be brought against the other eight administrators whose cases were dismissed.[43] Only time will tell if Nessel's new investigation will deliver these criminal charges and convictions—or if this development is just another stalling tactic designed to placate Flint residents.

The Legal Situation

In recent years, Judge Judith Levy has hosted meetings regarding legal concerns connected to the Flint water crisis. Typically, over 100 people show up. They are mostly attorneys involved with the Flint water cases.

In response to the evidence against water regulators and public administrators, personal injury and civil rights attorneys descended upon Flint, invading homes through television commercials and cars through radio ads, all attempting to entice residents with the promise of civil judgments. The attorneys represent parties involved in at least ten class action lawsuits and over fifty individual actions in response to the crisis.[44]

After documents implicating state and federal agencies surfaced, Flint residents began filing their own lawsuits. On July 26, 2017, the first status conference among attorneys engaged in the class action lawsuits and individual claims against government officials and engineering firms took place.

Some attorneys who participated—such as Esther Berezofsky of Williams Cuker Berezofsky, LLC—had been incredibly successful in encouraging residents to join class actions. This firm represented plaintiffs in the Lowery class complaint, as well as over 3,000 individual plaintiffs in cases concerning the water crisis. Hunter Shkolnik from Napoli Shkolnik also represented over 2,500 Flint families, while Corey Stern represented 2,027 individual plaintiffs.[45]

Attorneys aggressively pursued Flint residents. Firms that took on a large number of clients, like Shkolnik, employed tactics like enlisting Black actor and Harvard Law graduate Hill Harper to appeal to the population. Dressed in black in a local commercial—with background images from the civil rights era and the Flint water crisis—Harper intoned that he represented a firm called NorthStar, a law group he had founded to pursue social justice. In the ad, he encouraged Flint residents to give his organization a call to discuss their legal options: "I'm worried about the families of Flint, Michigan [since] there is so much stress and confusion around the Flint water crisis lawsuits."

NorthStar's phone number was prominently displayed on the right-hand side of the screen. During the first twenty-three seconds of this one-minute commercial, references to the firm that he was *actually* soliciting for—Napoli Shkolnik, which had contracted him to attract Black clients—were barely visible (in small black font in the darkened right-hand corner of the television screen). Napoli Shkolnik's name shows up in the closing disclaimer of the commercial, just as Harper quickly concludes his message to Flint: "The first set of lawsuits will be closing soon. So please don't get left out. Call us."

Given the number of pending cases, Judge Levy consolidated the class actions and individual cases to streamline the process. Two sets of liaison counsel were assigned, and Levy ordered these attorneys to file amended master class actions to consolidate pending cases. Although attorneys for government officials collectively opposed the consolidation on the grounds that the motion was premature, it moved forward nonetheless (with, of course, conflict).

One significant hurdle to the consolidated cases was the resolution of immunity claims by government defendants. Plaintiffs' attorneys pushed for discovery, while government attorneys strongly opposed what they called proceeding "willy-nilly with discovery . . . without really keeping an eye on what is also occurring in the State cases."[46] The government persuasively argued for delaying discovery, particularly in cases like those against Eden Wells and Nick Lyon, who were facing criminal charges in state court. Another set of cases tangled with immunity and other vital legal issues that attorneys dubbed "The Big Six," which included Village Shores, Alexander, Washington, Gulla, Walters, and McMillian. These cases had not been adjudicated. Conflicts also emerged among plaintiffs' attorneys. Interim co-lead class counsels Theodore Leopold and Michael Pitt and co-liaison counsels Hunter Shkolnik and Corey Stern argued over fee arrangements and decried a lack of meaningful collaboration. Leopold and Pitt wanted to determine the class and common benefit fees as well as compensation before settling the litigation, while Shkolnik and Stern believed that it was inappropriate to discuss compensation until the litigation was resolved.

Leopold filed motion to replace Shkolnik and Stern in March 2017, a motion Shkolnik characterized as "nothing more than a blatant money grab . . . that is a compilation of speculation, hyperbole, innuendo, and in some places outright falsities."[47] Shkolnik accused Leopold and Pitt of "side fee deals" that involved delegating work assignments and hours to firms with which they had fee-sharing agreements, thus increasing their potential profit from the Flint water litigation. Shkolnik noted that he had flagged this issue in July 2017 and proposed a time and expense order upon learning that Leopold was sending teams of attorneys out to "supposedly 'map' lead poisoning," but instead, he claimed, "these teams were actually engaged in client solicitation, not mapping."[48] Leopold and Pitt would not agree to a time and expense order unless Shkolnik agreed to stop accepting new clients and to allow Leopold to receive 80 percent of the common benefit funds and one-third of attorneys' fees from the individual personal injury cases.[49] With what we can assume was no small

measure of dismay, the judge encouraged the attorneys to get along and advocate for their clients: the people of Flint.

Engineering Services Join the Legal Fray

There was another set of attorneys who voiced distinctive concerns during this
time: those representing the group called the "engineering defendants." These
attorneys persistently requested that the plaintiff class (including adults, mi-
nors, and decedents) produce fact sheets detailing plaintiffs' personal injuries,
property damage, and business losses. In addition to requesting assigned re-
leases for medical, employment, education, disability, and insurance records,
defendants consistently asked the court for access to plaintiffs' address information, including how long they had lived at the residence, whether the home
had lead pipes, and their blood lead level results, if available.[50]

Despite these requests, attorneys for the plaintiffs had been slow to comply with these records requests. As attorney John Grunert, representing the
Veolia North America (VNA) engineering defendants, complained, "The
VNA defendants were in Flint for a month. And we've turned over more than
12,000 internal documents. LAN [Lockwood, Andrews, and Newman, Inc.]
has turned over more than 80,000. The state has turned over hundreds of thousands. [But] the plaintiffs in these cases have given us nothing about their cases.
Zero. Zip."[51]

Flint retained engineering firms Lockwood, Andrews, and Newman, Inc.
and VNA to evaluate its water distribution system so as to verify the city's
compliance with state and federal environmental regulations. They were also
charged with making recommendations to improve water quality. LAN had
been hired by emergency manager Ed Kurtz in June 2013 to help determine the
best course of action in order to use the Flint River as the city's primary drinking water source while the Karegnondi Water Authority pipeline construction
was underway. LAN continued to advise regarding the city's transition to Flint
River water through 2015. It was paid over $3.8 million for its expertise and
recommendations.

VNA came on board around mid-February 2015 to continue advising about
the transition to river water. According to the plaintiffs, however, VNA failed
to relate corrosion to high total trihalomethanes in its "Flint Michigan Water
Quality Report." Attorneys representing the plaintiffs also maintained that
LAN's report titled "Trihalomethane Formation Concern" overlooked corrosion as being a root cause for Flint's escalated TTHM and claimed that these
firms failed to provide reasonable engineering expertise, thus contributing to
the destruction of the water distribution system that they had been hired to
help improve.

"Had either LAN or Veolia performed such an analysis," the plaintiffs argued, "it would have quickly revealed that Flint River water was contaminated by corrosive salt accumulated . . . and revealed that the City of Flint had not adapted a corrosion control protocol as mandated by the federal Safe Drinking Water Act and related Lead and Copper Rule." Further, the plaintiff class maintained that these companies had the capacity to help Flint avoid corrosion problems but neglected to address "a series of red flags," including Flint's seemingly losing battle with coliform bacteria, the legionella outbreak, and "the very color of Flint's tap water, which was rusty and brown precisely because it was leaching metal from Flint's pipes."[52]

Government Officials as Defendants

The emails between state regulators, obtained via FOIA requests, certainly made these bureaucrats look bad. But most of the blame for the Flint water crisis has been laid at the feet of former governor Rick Snyder, even though his attorneys contended that, on the contrary, there was no evidence connecting him to the decision-making process in Flint. Richard Kohl, arguing on behalf of Snyder in his role with the Michigan attorney general, maintained that, when it came to the water source switch, the governor had no involvement.

"There's no reason for the governor or anybody else to believe that not to be true. There was no reason for the governor to not rely on Flint, not rely upon its trained professionals in the Department of Environmental Quality. This was their area of expertise. He's not an environmental engineer," Kohl argued. "There's no reason he shouldn't be able to rely upon them to make those decisions and follow through to ensure that that happened."[53]

Kohl went on to lay blame for approving the use of the Flint River as a public water source—without making necessary repairs and upgrades to the Flint Water Treatment Plant—at the feet of "the City of Flint and its consultants," who should have been "identify[ing] what needed to be done." Kohl directed further blame toward the DEQ (which approved the application to operate the treatment plant) and its employees, specifically Stephen Busch, the Lansing water supervisor for the DEQ's Office of Drinking Water and Municipal Assistance, and Pat Cook, the DEQ water specialist who signed the limited permit as proxy for Mike Prysby.[54]

Plaintiff attorney Renner Walker argued that, as the complaints started rolling in, Snyder became aware that people were getting sick and failed to "hee[d] to a call to action."[55] Not only did the governor authorize the Karegnondi Water Authority—against advice from an independent firm that found that staying with the Detroit Water and Sewerage Department made more economic sense—but he prolonged Flint's exposure to contaminated water.

Overall, attorneys for the plaintiff class claimed that the government defendants covered up and extended the exposure of Flint residents to corrosion and lead-contaminated water. For instance, in addition to falsifying test results, the plaintiffs maintained that the DEQ rigged Lead and Copper Rule test sampling procedures (essentially, tampering with evidence). The department further misled the EPA about Flint's corrosion control procedures and misrepresented the cause of lead issues in a resident's home (which it ascribed to "lead sources within the home," even though it had plastic piping).

Other assertions included that the government defendants failed to provide adequate warning regarding the risks of consuming Flint's drinking water, especially after it had become clear that the city was dealing with bacterial contamination, a legionella outbreak, *and* elevated lead levels among preschool children. Michigan health officials believed that the Legionnaires' outbreak was the worst in the state's history, yet the government defendants neglected to inform Flint residents about the danger stemming from consuming city water.

In other words, government officials appeared to undermine efforts to identify the health effects of changing the source of water for Flint. Well before Dr. Mona Hanna-Attisha analyzed Hurley Medical Center data, she requested blood lead level data from the MDHHS. She was worried about children's lead levels after the water source switch, informing the data manager for the Healthy Homes and Lead Poisoning Prevention programs within MDHHS, Bob Scott, that her results indicated there was a serious problem. But she did not receive any support or request for collaboration.

Instead of working to support the children of Flint, MDHHS officials began to email each other about how best to undermine Hanna-Attisha's findings before she could even present her data. In an email to various state employees, Governor Snyder's press secretary said, "Team, [h]ere's the data that will be presented at the Hurley Hospital press conference at 3 P.M. As you'll see, they are pointing to individual children, a very emotional approach. Our challenge will be to show how our state data is different from what the hospital and the coalition members are presenting today."[56]

After the study came out, Scott advised MDHHS to "say something like this: While the trend for Michigan as a whole has shown a steady decrease in lead poisoning year by year, smaller areas such as Flint have their bumps from year to year while still trending downward overall." In an email discussion between Scott and MDHHS spokesperson Angela Minicuci, we also learn that Nancy Peeler, then director of MDHHS's Program for Maternal, Infant, and Early Childhood Home Visiting, mentioned that "my secret hope is that we can work in the fact that this pattern is similar to the recent past."[57]

Once high lead levels were exposed—both by community researchers and by those from Virginia Tech—admissions of wrongdoing began to emerge. After Flint returned to the Detroit water system, DEQ director Dan Wyant acknowledged in an email to Governor Snyder and other government officials that the department had botched Flint's water distribution by failing to implement corrosion control protocols: "I believe now we made a mistake. For communities with a population above 50,000 optimized corrosion control should have been required from the beginning."[58]

A Work in Progress, Not Progressing

By February 2019, over a year and a half after attorneys initially met on behalf of the Flint water crisis parties, they were still negotiating the parameters for discovery.[59] Despite the slowdown in the process, attorneys continued to tell Flint residents that the cases were headed in the right direction, ensuring them that they would be eventually compensated for their losses and distress.

During a class action suit update meeting held at United Auto Workers Local 659 on June 27, 2019, about 100 Flint residents showed up to listen to Ted Leopold and his co-counsel, Michael Pitt (recently assigned to the class action by Judge Levy), provide an update on the Flint water crisis civil litigation. Both attorneys declared their cases could still be won, despite the fact that the new attorney general had recently dropped the criminal cases against the remaining defendants. Leopold told the Flint crowd he understood that they still had concerns while also letting them know how clear it was that their urge to continue fighting for justice remained intact.

"I see your T-shirts, I hear you say that Flint is still broken," Leopold said. "We could talk about how Flint is still broken, with its pipes, its illnesses, its bad water, but what is not broken is your spirit . . . and [commitment to] making sure that what happened to you will never happen again." Leopold told the crowd that as their attorneys, they had won almost "every aspect of this litigation . . . because justice always wins in the end." Leopold and his co-counsel also announced to residents that they would be pursuing options for settlement in order "to see if there are ways that can formulate a win-win for everybody."[60]

But Flint residents, including Claire McClinton, were up front with their criticism, describing how they would be slow to accept empty promises and advice from anyone attempting to tell them how they needed to shape their social justice response to drinking water injustice. As McClinton put it, when at the podium in front of her fellow residents who were still emotionally raw about the water crisis, "We appreciate the attorneys and the support they're giving us navigating the court system, we appreciate the scientists, the Christian community, the charitable community doing the job the government should be doing, and we appreciate our representatives speaking out in the hearings. But in the final analysis . . . we the people will decide when justice is done. None of these support groups will tell us. We will say when justice is served. We're the ones making that decision."[61]

Despite the admissions of wrongdoing, victories, and cooperation secured through the legal process, circumstances remained largely unchanged in Flint after President Obama came to town and personal injury and civil rights attorneys made promises to the city's residents. Although civil litigation provided a path to recovery and restitution that eventually led to a $600 million settlement offer from the State of Michigan in August 2020,[62] litigation efforts against parties deemed responsible for the water crisis, including the EPA, continue to be a work in progress.

Throughout Flint's journey with water crisis litigation, its water distribution system was still compromised. As Flint's plaintiffs maintained, the infrastructure running under their streets and into their homes had been corroded "to a point where the only viable means for ensuring the transportation of safe water into the homes . . . was replacing the pipes."[63] Until *all* the possible sources of corrosion and lead are addressed, many people in Flint will never trust the water. And for good reason.

7

Crisis as a Daily Grind

Digging Up Water Pipes
Whether or Not They Need to Be

It was standing room only in Flint's city hall on Monday, November 9, 2015, when Dr. Karen Williams Weaver was sworn in as the new mayor. With cameras flashing, reporters, city officials, and hundreds of Weaver's campaign supporters and family piled into the city council's chambers and spilled out into the hallway to witness the swearing in of the city's first Black woman mayor.

Weaver, a clinical psychologist who was active on the boards of many local philanthropic organizations, hails from a family of community leaders. "Service is in my DNA," she proudly announced to the crowd. Service, according to Weaver, "is not a job. It's a calling."[1]

Her mother, Marion Coates Williams, became Flint's first Black classroom teacher in 1943 after earning her bachelor's degree (in two years) from Eastern Michigan University and then a master's degree in elementary education from the University of Michigan. The mayor's father, Dr. T. Wendell Williams, was one of Flint's first Black pediatricians; in 1963 he became the first Black person on Flint's board of education via an election that pitted eleven candidates for the three vacant spots on the board.[2]

Thanking her family and volunteers for their support—as well as water activists who "stayed vigilant in the fight for clean and affordable drinking water"—Weaver told the crowd that her choice to run for mayor "was not a whimsical decision." Despite backing from prominent clergy, she related that many considered her run as a passion project that would not be successful, given that the odds of defeating Mayor Dayne Walling were long. "According to the experts, we were defeated before the votes were even cast. [But] they underestimated us. . . . Friends, we stepped out on faith. Faith in ourselves. Faith in each other. Faith in our collective resolve to bring the need of change and to give Flint residents the democracy we deserve," she preached.[3] With supporters in the crowd urging her on, Weaver pledged to work diligently and give Flint residents "a handsome return on their investment." Her next steps were straightforward:

> On November 3, residents said in a loud, clear voice that they were sick and tired of poisoned water. They said that they were sick and tired of high water bills. They said that they were sick and tired of being undervalued and marginalized. Sick and tired of double-talk and political-speak. And they said over and over again that they want real change.... They said, we want to charter a new course.... At this moment we celebrate a victory for all of Flint.... We didn't work this hard to preside over the status quo.... We ran on a platform of change.... But, let's be real. If it's a continuation of business as usual, then this [will be] a hollow victory.[4]

Despite the excitement surrounding Karen Weaver's successful bid to be Flint's next mayor, the State of Michigan was still in firm control of her city. This was true even though the emergency manager had been removed in April 2015. By requiring the city, while under home rule, to seek guidance from a Receivership Transition Advisory Board in matters pertaining to "major financial and policy decisions," the state still held the purse strings.[5]

Relatedly, although critical issues with the city's infrastructure remained, there was limited agreement between the state and the city regarding how best to fix all the problems caused by the Flint water crisis. As a result, the city's quest for water equity continued to be an uphill battle. This chapter documents Mayor Weaver's attempts to lead Flint during the initial phases of disaster recovery and pipeline repairs.

Weaver's Journey: From Rising Star to Scapegoat

Encouraging developments began to emerge after Weaver became Flint's mayor. After barely a month in office, she declared a state of emergency on December 14, 2015.[6] After that, the state and federal governments and private agencies began taking a more direct intervention approach, with the National Guard, American Red Cross, and other emergency response teams deployed to Flint from around the country.

By April 2016, these teams had visited 21,291 homes and more than 5,000 apartments to distribute 84,505 water filters, 24,866 water test kits, 195,264 filter cartridges, and 153,005 cases of water.[7] Points of distribution were also set up by state and city agencies at churches and fire stations, enabling residents to pick up these items. In addition, the state maintained that all residents would be provided one filter and 211 hotline information to help the elderly and disabled arrange a one-time delivery of bottled water and filters. By mid-September 2016, almost 3 million cases of water had been distributed to Flint residents.[8]

To make sure that resources provided to residents were properly installed and maintained, the City of Flint created programs like CORE (Community Outreach Resident Education). CORE staff visited residences to provide water filter installation instructions and guidance concerning water use and resources for medical services.[9] But, just after the federal Environmental Protection Agency awarded Michigan's Department of Environmental Quality a $100 million grant to repair Flint's water infrastructure in March 2017,[10] help for the city began to dry up.

Around this time, politicians and administrators at the state level became preoccupied with self-preservation and protecting themselves from civil and criminal charges. By April 2017, the state had reimbursed the legal fees of employees of the DEQ, the Michigan Department of Health and Human Services, the Michigan Department of Treasury, and the City of Flint. Over one-third of these funds were used to defend Governor Rick Snyder, who hoped that immunity would shield him from legal claims stemming from the crisis.[11]

State and local administrators and water regulators apparently needed the legal advice, since in mid-June 2017 the Michigan Department of the Attorney General had announced its intent to charge several state officials with involuntary manslaughter stemming from their role in mismanaging the Legionnaires' outbreak (which took the lives of, at a minimum, twelve people). In October in that year, the US Committee on Oversight and Government Reform sent a letter to Snyder asking him to clarify his sworn testimony in which he claimed that he had not learned about the Legionnaires' outbreak until 2016.

By late 2017, the excitement surrounding Mayor Weaver and her agenda had begun to sour. Residents were losing patience; Flint's water was still unsafe. The official directive from the Weaver administration was to continue drinking bottled water until at least late 2019.[12] The city had returned to using Detroit water, but damage to the lead pipes was extensive: 55,000 service lines were slated for replacement, with only 4,535 having been fixed as of September 30, 2017.[13]

Repairing the damage to Flint's water infrastructure would be an expensive endeavor. Residents found out that the cost was higher than the $55 million Weaver initially estimated. In fact, according to a 2016 state report, the total price tag—which eventually included fixing damaged pipes, safely disposing of lead-contaminated soil, and supplying residents with bottled water during the project—topped $216 million.[14] By February 2016, Weaver admitted that it would take over a billion dollars over the next decade to completely deal with what the water crisis had done to Flint.[15]

Around the same time, Governor Snyder said that removing the pipes in

Flint wasn't on his "short-term" agenda. "It's a lot of work to take out pipes, to redo all of the infrastructure, that's a whole planning process," Snyder declared. Instead, he announced that the state's primary strategy for resolving Flint's lead problem involved coating corroded pipes with phosphates in order to prevent lead from leaching into the water.[16]

Despite the state's role in creating the water crisis, its primary strategy was curiously underwhelming, a Band-Aid approach if ever there was one. The absence of a pipe removal plan on Snyder's "short-term" agenda meant that Flint would need to wait a good long while before a permanent solution was carried out. The Michigan legislature did vote on a $28 million emergency spending bill for Flint, but that was inadequate for the actual situation on the ground and was a devastating blow to efforts by activists and citizens to push the issue into the national spotlight.

Meanwhile, the City of Flint began expressing its concern over unpaid bills from residents for water they did not trust (in part because they had been told to avoid using it without a certified filter). Both residents and the city council were told by city administrator Natasha Henderson that Flint's water utility was in a "very precarious situation" and could go under by the end of the year if residents continued to withhold payment for their lead-tainted water.[17]

A few Flint residents responded with a water shutoff protest outside of city hall in late January 2016. Over fifty residents chanted the familiar "Hey, hey-ho, Snyder's got to go!" and "Don't pay the bill; the water will kill." Residents complained that they couldn't pay the high bills—which were some of the highest rates in the nation—for incredibly poor-quality water.

"That's just wrong," Melissa Mays told a news reporter covering this protest. "There's nothing that makes sense in any of this. . . . They just keep billing us and billing us. This just needs to stop!"[18]

Residents—who had had enough of the back-and-forth—opted to take matters into their own hands. Given that the city and state wouldn't agree to carry out pipeline replacement, the Concerned Pastors for Social Action in Flint— which consisted of about thirty prominent church leaders—and Mays, with the help of the Michigan ACLU and the Natural Resources Defense Council, filed a lawsuit in federal court against state and city officials in January 2016. They sought a court order that would require both parties to replace lead pipes, set a timeline to complete the job, and make provisions to ensure that the water provided to Flint's citizens was safe.[19] In a March 2017 settlement, the City of Flint agreed to replace 18,000 lead or galvanized service lines by 2020 (with the help of up to $100 million from the state of Michigan and funds set aside by the US Congress).[20]

"This is the first little battle won in this huge overall war," Mays told a reporter. "For the first time, we've been able to have a federal court enforce the state to do the right thing, which is to replace the pipes that their agencies and their administration broke."[21]

But before seeing real rewards from their latest victory, the community that had been through so much continued to experience setbacks. By June 2017, the state had closed ten of the fourteen water point of distribution sites and eleven of fourteen help centers. Point of distribution sites were where residents had been able to gain information about food (especially produce) that helped lower lead absorption in the body, health care options, and mental health services, along with bottled water and replacement cartridges for water filtration systems.[22]

Flint's people felt understandably abandoned. Citizens expected Weaver to stand up to Governor Snyder and demand free bottled water until the city's pipe replacement process was complete. But, less than a year into her term, residents were losing access to free water.

By May 2016 the mayor was in the hot seat because of accusations that she moved money from city water funds to her own political action committee. The now former city administrator Natasha Henderson—who had served in this capacity since 2014, working under emergency manager Darnell Earley— had filed a whistleblower lawsuit against the mayor and the City of Flint for wrongful termination. She alleged that she had learned from Weaver's assistant that the mayor had been diverting money—set aside for Flint's Safe Water Safe Home fund—to a campaign fund under her discretion. Henderson claimed that Weaver illegally fired her three days after asking the city attorney to investigate the matter. Declaring that Henderson's allegations were "outrageous" and "completely false," Weaver maintained that Henderson was fired due to her failure to inform the city of the Legionnaires' disease outbreak, which Weaver claimed Henderson had been aware of since March 2015.[23]

District Court judge Sean F. Cox initially dismissed Henderson's case because her claim of whistleblower protection was not valid under the First Amendment due to her status as a city employee (as opposed to a private citizen).[24] An appeals court later reinstated Henderson's case, but she eventually lost at trial when no evidence of financial mismanagement emerged to support her claims, which led the jury to decide she was not fired due to any discovery of unethical behavior on the part of Weaver.[25]

Weaver was vindicated in court, but the damage to her reputation was done. Some residents who felt hopeful with the election of Weaver as the city's new mayor began to worry that, like the state, her commitment to getting the funds and resources necessary to resolve the water crisis had wavered. Their worries and disappointment intensified once the city, under Weaver, began actively pursuing delinquent water accounts.

Residents' frustrations with their water bill accounts escalated during the public comment period at a well-attended May 18, 2017, city council meeting that took place just a few weeks after over 8,000 residents received notice from the city that they would need to settle their water bill or face a tax lien and potential foreclosure. At this council meeting, there were plenty of residents present who were directly impacted by the city's decision to go after delinquent water accounts.[26]

One by one, residents stepped to the podium to share their stories. On the whole, they tended to agree that they should not be penalized for failing to pay for contaminated water. As one resident noted, "I'm one of the water customers that received a tax lien on my property. I refuse to pay for poisonous water back in 2014, 2015 and 2016. So yes, $1,800 has been placed on my property with the threat of losing my property, my home for this."

Residents like local civic leader and frequent city council attendee Quincy Murphy argued that the city was forcing residents to pay for defective goods and claimed that he wouldn't dream of giving his money away to them: "If I go to the store and I get a bad product, I'm going to go and get my money back. The city of Flint has not given the residents any kind of money back, a credit.... We want our money back for this water we had to pay for that we can't drink and bathe in."

From some residents' perspective, their obligation to pay for contaminated water should have been addressed by Mayor Weaver, who they believed could end the exploitation by overlooking or revising the ordinance that required Flint residents to pay for city water. Instead, Weaver appeared to have reversed her position on paying for the city's water after being elected mayor. One local, Arthur Woodson, a Black community activist and Flint resident, said at this May 2017 city council meeting,

> Before the mayor got into office, she advocated "no one pay your bills." The mayor didn't even pay her bill in 2013; she just paid in October of last year. How are you going to advocate that nobody pay their bills over a year and a half [until] you get the job that the people put you in paying $90,000? You want to shut off people's water and put tax liens on their houses where they still can't even use the water. We are worse off now than we were back then. We didn't have shutoffs, [and] we didn't have tax liens.

After managing to fend off Henderson's allegations and the persistent complaints about the cost of tainted water, the hits just kept coming for Flint's new mayor. A year after the filing of the Henderson lawsuit, residents reached such a point of frustration with unresolved water quality and cost issues that, in August 2017, they gathered enough signatures for a recall election. Arthur Woodson spearheaded the successful petition drive that aimed to oust Weaver via a recall election in November. The recall petition was approved by the Genesee County Board of Electors soon thereafter.

Folks like Woodson wanted Weaver out for various reasons beyond the accusations lodged by Henderson. He believed that the city was "worse off than [it] was before" Weaver had taken office. Her critics didn't like how she was running the city, claiming that she handled business too much like former mayor Dayne Walling, whose style was believed to have cost him his job.[27]

Woodson and other Weaver critics—who included several members of the city council, including Ninth Ward councilman Scott Kincaid—were especially critical of her push to contract with Rizzo Environmental Services for the city's trash removal over the current contractor, Republic. Weaver's critics made sure it was not forgotten that she continued fighting the renewal of Republic's contract until Rizzo was implicated, but not charged, in a federal corruption case.[28]

In addition to having concerns about business relationships, Woodson's chief complaint seemed to be that Mayor Weaver simply had not done enough for Flint. As Woodson mentioned at a September 11, 2017, city council meeting,

> Mayor Karen Weaver was in charge of the recovery. . . . But she wasn't even in the office five months before she started trying to divert the funds that she was going out to get for us. . . . After that, she endorsed Hillary Clinton. Turned the water crisis into a political football: it was the Republicans versus Democrats. . . . Then after that, she fought against the moratorium that you all passed that put out 8,000 liens. So let's stop talking about what she has . . . and [hasn't] done, because we still haven't seen her do anything yet.

Woodson was especially critical of the city's recovery from the water crisis. He was tired of the double-talk: "They keep on [telling] the public that [the service line replacement is] almost complete. [But] it's not," he said. Woodson contended that residents had made the efforts to bring attention to the water crisis and that the mayor was only benefiting from their hard work. It was the residents, Woodson claimed, who secured the "notoriety . . . when the lead experts came in and switched Flint back on Detroit water." He added that before Mayor Weaver was elected, "the media was already there," and that the people of Flint were the ones who "brought the money . . . and the government here. It wasn't Mayor Karen Weaver [who] did that." In the end, Woodson declared that Weaver needed to do her part because lives were on the line: "Water is the major issue . . . but we're still not able to drink from the tap. . . . So let's not get it twisted. . . . People are dying and Mayor Karen Weaver [is] still not doing her job."[29]

Despite the criticisms voiced by Woodson, Kincaid, and sixteen other candidates for the position (including the charismatic, shovel-carrying Chris Del Morone, a frequent speaker at city council meetings), Weaver eventually won the city's November recall election in a landslide.[30] But tensions remained high, especially between the mayor and several members of the city council. Trust had been broken.

Relationships were so damaged that lines of decorum were crossed and name calling commenced. Council members like Kate Fields publicly accused the mayor of corruption, while Kincaid claimed the mayor was incompetent.[31] Weaver also had a tough time getting members of her team approved by the council, including Hughey Newsome as the new chief financial officer. He would eventually secure the position but ended up offering his letter of resignation after a short stint, complaining of the city council's consistent "lies" and "innuendos" that Newsome believed were shared with the press in an "attempt to slander [his] professional name."[32]

Ultimately, Flint residents and some members of its city council didn't trust Weaver or members of her administration. According to Jackie Poplar, then city council member from Ward 2, Weaver tricked the Flint community into believing that she was ready to lead. Residents viewed her as a charismatic leader who had emerged during the crisis, but, according to Poplar, the mayor had not "done enough since she got in office" and, at the May 18, 2017, council meeting, said, "People got hoodwinked because of the water crisis. This vote to get her in here was on people's emotions. If a busted rat ran for mayor, that rat would've probably won because of people's emotions. People voted thinking she was Jesus' daughter, finna [sic] turn water into wine, not so."

But, according to Newsome in his March 2019 letter of resignation, Weaver was never provided the leeway and cooperation from the city council that she needed in order to govern Flint effectively. He faulted the city council's drama and infighting, as well as members' tendency to "weapon[ize] the press to further divide an already divided community." His two-page resignation letter urged the city to move beyond "politics as usual" because, from his perspective, this behavior was undermining the work they needed to carry out in service of Flint residents:

> Some of you view the world through a lens of political one-upmanship. Without any trepidation, you move outside of your elected positions as legislators to "witch-hunt" the Mayor, her administration or one another.

Please remember that while the Mayor is not perfect, she and her administration work tirelessly for this city and this city is entrenched in our hearts. I can attest to the countless hours Mayor Weaver and her staff put into keeping this city running. Somehow that escapes the dialogue that happens in Council chambers on a regular basis. Council also goes out of its way to attack one another day in and day out, preventing productive dialogue and respectful debate on incredibly important matters.[33]

The distrust and disrespect between city council members and Weaver's core staff peaked after Newsome vacated the chief financial officer position. Not even six months after he left, city council members were accusing Weaver and her staff—at a meeting on September 4, 2019—of boycotting meetings and avoiding direct questions about contracts and commitments made during Weaver's tenure.[34]

In response, those on Weaver's staff—including the new chief financial officer, Tamar Lewis—complained that council members occasionally spoke to them in a condescending manner. During this September city council meeting, before abruptly walking out during a discussion on the status of the city's water fund and the water liens, Lewis told the council she didn't "need all of this scolding. Just ask me the question."[35]

In addition to being aggressive with how they communicated with Weaver's staff, council members' behavior led some city workers to complain that the council's requests were often inappropriate. Issues included requests not being communicated in writing and some being listed on the calendar for discussion only moments before a meeting began.

But community activists like Woodson countered by insisting that residents were stuck in the middle of this toxic chaos. He said that the city workers are just as "bad" as city council members, reminding the council about "when Tamar Lewis called me an 'asshole' and a 'cock sucker,' didn't none of ya'll say anything about that. None of ya'll."[36]

These ongoing conflicts between Flint's stakeholders have certainly caused significant setbacks among community activists, city leaders, and city staff. Microaggressions and power struggles—not to mention the occasional name calling—have greatly undermined the pace and scope of the policy implementation necessary to help Flint get beyond its water crisis. The conflicts have only served to distract leaders and activists from their shared and primary objectives—helping the city. However, these conflicts are not the true source of the delays that have plagued Flint's recovery.

In fact, the State of Michigan's peculiar and tenacious tendency to govern areas like Flint with limited regard for residents' well-being and property rights is the chief reason the city's recovery has been so sluggish. Instead of systematically and directly resolving issues that stemmed from years of ne-glecting the city's water infrastructure, the state—in tandem with associated local and federal agencies—worked to mitigate the costs of solving the crisis. Government administrators and water regulators made attempts to restrict how much to spend on Flint's recovery by prematurely declaring water safe to consume while also seriously compromising and underfunding the needed pipe replacement program.

Saying the Problem Is Solved Doesn't Solve the Problem: The Rush to Declare Victory

It had been four years since Flint River water had started flowing from Flint's taps when, on April 11, 2018, city residents took their protestations about persis-tent water problems to the Capitol Building in Lansing. They were exhausted by years of high bills, water tests, petitions, protests, lawsuits, and physical ail-ments. Even though over $350 million in private donations and federal funds had been harnessed to aid them with their water crisis, it was still there, with no real resolution in sight.

The last of the four state-run bottled water distribution stations closed April 6, 2018, as the state declared Flint's lead and copper levels had been below fed-eral limits for almost two years. Apparently, that meant the water crisis had been resolved. The EPA claimed that 90 percent of Flint water samples had less than 4 parts per billion of lead over the last six months of 2018, the low-est level of contamination since the start of the crisis.[37]

Disappointed by Michigan's decision to shut down the water distribution stations, Dr. Mona Hanna-Attisha—the Hurley pediatrician who had been crucial in bringing notice of the rising blood lead levels among preschool chil-dren in Flint—registered her complaints by tweeting, "This is wrong. Until all lead pipes are replaced, state should make available bottled water and filters to Flint residents."[38]

Although the State of Michigan and the EPA declared the water safe to consume, research showed that water problems were likely to be an ongoing concern throughout the city for some time.[39] In a 2018 article in *Environmental Science and Technology*, Kelsey J. Pieper and colleagues pointed out that several homes in Flint experienced sporadic spikes in particulate lead throughout 2016. They concluded that testing in these homes produced inconsistent results, even between different taps within the same house. Particulate lead moves unpre-dictably within water distribution systems, and the well-documented plumb-ing corrosion of soldered joints and brass fittings in Flint's homes ensured that lead would continue to plague residents.[40] Researchers simply disagreed with state government and EPA claims that the water problems were resolved. On the contrary, researchers warned that Flint's water quality woes would fluctuate and that water treatment experts should be prepared to respond to this reality.

Perhaps the truth is that the State of Michigan was just looking for a way to finally end its water intervention in Flint. And, despite the potential consequences of premature assurances, it would use its safe-water declaration to legitimize the reestablishment of inconsistent development policies and poor treatment of Flint's citizens.

Mayor Weaver vehemently disagreed with initial attempts to declare the water clean in 2018. She did so again when assurances were repeated on June 3, 2019, by EPA acting director Andrew Wheeler (a few months after he was appointed to this position).[41] Weaver pointed out that she would not declare the city's water safe solely on EPA assessments: "This feels like another attempt to rush this water crisis into 'being fixed' or 'being solved,' [but] the reality is that this is a process. Not only is it a process, but it is the first of its kind.... The medical community and scientific community will both have to be in agreement, after a period of testing over time, that the water is safe to drink before I ever declare it safe."[42]

Residents at Flint City Council meetings regularly expressed concerns with the state and federal government's premature assurances about the quality of their water. At a November 29, 2018, meeting, Suzanne Green talked about her frustrations with federal and state declarations, given the water quality issues she continued to have: "I don't know why any of this travesty is going on in this community. However, I am here to tell you, if you people for one second think this water is safe by any means, you're wrong! It is damaging my grandbaby. She already got damaged from the lead; now she's getting damaged from the chlorine. The city needs to do something. My home, I can't sell it, I can't rent it, I can't do nothing. It's a death trap."[43] In addition to being upset about the state's willingness to overlook existing water problems, detracting from the public narrative that Flint's water now met all federal standards, Flint residents' ultimate concern seemed to be their belief that the water problems would never be fixed.

A FAST Start Slows Down

This feeling was lingering more than two years after the city had launched FAST Start, its pipe replacement program, in March 2016. It has had a sobering and depressing effect on a segment of Flint's population, especially those whose water problems and related issues have not been resolved. As one person stated

Figure 7.1. Flint's FAST Start process breaks ground.

at the November 29, 2018, city council meeting, "The city of Flint doesn't have [the solution]. I just wanted to come as a local business owner. I reside in Vegas; but my business will always be here. We pay taxes here. [But] I'm not up here for me. . . . A lot of my friends are here [and] relatives are here. When I come back home, what I see on most of my friends' and family's faces, is hopelessness. It's hopelessness."[44]

As part of water quality recovery efforts, the City of Flint had launched the FAST Start program and was required to complete the four phases of this process by 2020, according to the Concerned Pastors for Social Action settlement agreement. Phases 1 to 3 received $25 million from the state in 2017. The program was also supported by $100 million from the federal Water Infrastructure Improvement for the Nation Act of 2016, which was aimed at assisting infrastructure improvements in Flint.[45]

During the initial stages of the process, the city's strategy was to get the lead out of Flint by repairing "30 lines in 30 days." Residents were warned that construction crews would not be working one street at a time. Instead, the intent was to use geographically dispersed sampling in order to go to high-risk locations first (based on water sampling in every neighborhood).[46]

After the sampling was completed, the bidding process began. By March 2017, six companies and two labor unions were awarded contracts by the city council. Most of the construction crews came from Flint-based companies, including Goyette Mechanical, which was responsible for 2,100 pipeline replacements. A "minority, woman-owned company," WT Stevens was assigned up to 2,700 water lines, and Waldorf & Sons, a Mt. Morris company, was assigned up to 600 lines. The companies that secured these contracts were tasked with completing 6,000 replacements by the end of 2017 (weather permitting). Lead or galvanized piping that ran from the water mains to home water meters was to be swapped out for copper piping with brass fittings to a new or existing water meter.[47] At the end of December, the city hired AECOM Engineering for $5 million to oversee contractors and crews.

"We are confident that we have a good team in place," declared Mayor Weaver in 2018. "When this project is complete, we want all Flint residents to know that the work has been done to ensure the water flowing from the tap is drinkable without the need for a filter. We have a team in place with the world-renowned expertise and experience needed to complete this work properly, efficiently, and on schedule."[48]

In addition to replacing lead and galvanized steel service pipes, AECOM was to address a range of water infrastructure improvement projects, some of which would not be completed until 2024. Senior program manager Alan Wong of AECOM stated, "AECOM is honored to have been selected by the City of Flint to manage the efforts to improve the City's water system. Our goal is to significantly raise the level of water quality throughout the City, for homeowners, residents, and businesses. We know this project is extremely important and the citizens of Flint expect their water system and water quality to meet all local, state and national standards, as it should."[49] The work AECOM would manage included repairing ground storage reservoirs and pump stations with structural issues, replacing water mains that were oversized for the demand and water meters that no longer worked, and installing an automated monitoring system that would provide instantaneous water quality updates.[50]

To assist Flint's efforts to locate hazardous service lines, researchers Jacob Abernethy from Georgia Tech and Eric Schwartz of the University of Michigan developed a predictive computer model. Starting in August 2016, it was expected to assist Michael McDaniel—a retired National Guard general who was overseeing the FAST Start program—in selecting high-priority addresses that had the highest probability of having lead or galvanized steel service lines.[51]

The model, updated throughout 2016 and 2017, was quite accurate. In September and December 2016, almost 90 percent of excavations revealed lead or galvanized steel lines. By May 2017, the city had provided the researchers with over 140,000 handwritten index cards that contained the work history of the city's entire water distribution system. Schwartz and Abernethy digitized about 50,000 of these records and, along with excavation findings received from McDaniel, used the data to continue fine-tuning Flint's pipe replacement efforts.

According to Schwartz, communication with the city and McDaniel was pretty consistent—until AECOM became involved. Early meetings at this stage included updates regarding previous work and discussions concerning the predictive model and how it would make the replacement efforts more cost effective. Schwartz and Abernethy were committed to continuing their support of the FAST Start program in collaboration with AECOM, but the company stopped returning their calls or correspondence. After turning over the rest of the digitized index cards to AECOM—as well as a list of 55,000 addresses indicating the probability of having lead or galvanized steel service lines— Schwartz and Abernethy realized that AECOM and the city had stopped engaging with them.[52]

Mayor Weaver and her team, however, would likely have argued that the city no longer needed Schwartz and Abernethy's model to guide their pipeline repairs since, at a press conference on December 4, 2018, she announced that "it's a day for celebration in Flint." According to the mayor, Flint's high-priority lead and iron pipes had been replaced, and, one year ahead of schedule, the city had satisfied its obligation under the *Concerned Pastors for Social Action* settlement to replace—by January 1, 2020—18,000 water pipes in areas with concentrated populations of vulnerable residents (such as mothers and children). Weaver announced that the city would now proceed with repairing the pipelines of residents with unpaid water bills.[53]

People weren't too quick to believe Weaver's declaration, since their observations of the pipeline replacement process made it plain that the city wasn't doing due diligence in locating lead and galvanized pipes. Residents felt misled by the mayor's announcement, because she misrepresented what the city was required to accomplish under the *Concerned Pastors for Social Action* settlement. Weaver made it sound as though the city had dug up more than 18,000 pipelines but neglected to mention that more than 80 percent of the work had been done to homes with *copper* pipes. The city had not meaningfully required that contractors target excavations in high-risk households that were more likely to have hazardous lead and steel pipes or children who had tested with elevated blood lead levels.[54]

In addition, while Flint residents were glad to see some progress, many thought they could no longer trust Weaver's discretion regarding AECOM after it was discovered she had received over $35,000 in campaign contributions from AECOM employees—including Senior Vice President Joseph Moss and project manager Alan Wong—and, additionally, PAC contributions from subcontractors that worked for the company.[55]

Then, after analyzing the city's efforts during the first phase of the FAST Start process, the Rowe Professional Services Company revealed to the *Flint Journal* that a third of the homes targeted by the city and Michigan's DEQ did not even have high lead levels.[56] By the end of May 2018, DEQ officials had formally inquired into why the city had paid WT Stevens and Waldorf & Sons almost $200,000 for completing excavations in homes that actually did not have lead pipelines.[57]

Residents had lost confidence in AECOM, too. The pace of the pipe replacement was glacial, yet AECOM requested more money before the end of the first contract year—to the tune of another $1.1 million to finish the job—while having checked the service lines of only 9,953 homes and replaced only about 15 percent of those.[58] Schwartz and Abernethy's predictive model had, from 2016 to 2018, boasted an 80 percent success rate in identifying lead water lines in need of replacement, but AECOM had abandoned that model. And the success rate for locating hazardous service lines had dropped precipitously—to just 15 percent.[59]

At a Flint City Council meeting, Ninth Ward councilwoman Eva Worthing asked representatives of AECOM why the predictive model for Phase 5 of the project was not being used. Vice President Moss retorted, "The model was never brought to our attention [and] that model was not discussed during negotiations. To assume someone's model's correct, that puts my firm at a tremendous amount of risk."[60]

Wong conceded that "the model is groundbreaking . . . but it [was] not practical to just use what the model said."[61] Phase 5, though, had been remarkably inefficient. The vast majority of the 7,000 addresses scheduled for excavation in 2018 were oversampled from areas where historical records indicated copper-to-copper service lines. Put another way, almost 75 percent of the work completed in Phase 5 took place in areas *known* to have service lines that *did not* need to be replaced.[62] It was an enormous waste of time and money.

No One Can Play Nice

Throughout this time period, the City of Flint and the State of Michigan were at odds. In 2017, for example, Michigan sued Flint for refusing to approve a long-term deal to purchase water from a Detroit-area system, claiming the city was threatening public health and safety by providing contaminated water. The DEQ found the city in violation of the Safe Drinking Water Act in 2017 and gave it 30 days to inform the state about planned corrective action (and 120 days to take such action). By 2018, attorneys for the state were back in court asking a federal judge to force the city to adequately respond to the water crisis as required by a lawsuit settlement.[63]

Also in 2018, the state sent the city an administrative consent order containing a timetable for bringing its water system back into compliance. Mayor Weaver claimed it was retribution for a lawsuit Flint had filed against the state for ending bottled water distribution. The DEQ countered that this was not true and that Flint had been in violation of the Safe Drinking Water Act for some time. In addition, the DEQ claimed the city was notified of its noncompliance well before its lawsuit against the state.[64] Clearly, effective cooperation was—and continued to be—unlikely.

By early 2019, the state was refusing to reimburse Flint for up to $97 million of expenses related to lead pipe detection and replacement efforts. State officials stated that Weaver, who gave AECOM the go-ahead for exploratory excavations, was wasting resources when there were far more effective lead-line detection methods available.[65] The city contended that it had been working to "get the lead out," wherever it happened to be, and was primarily using housing age as a guide for excavation efforts. Robert Bincsik, Flint's former water distribution supervisor and newly installed director of public works, claimed that the city had removed all hazardous service lines from densely populated areas in previous phases and did not expect to encounter additional lead or galvanized steel lines moving forward. The city, according to Bincsik, expected the rest of the areas to include newer properties with copper pipes, given that most of the homes in these areas were built in the 1960s and 1970s.

About the effort to find homes in need of pipe replacement, AECOM's Wong said,

> There is a screening process. You have to have an active account. And active accounts change every month here because some of the properties become a land bank property, which means they are not eligible to be part of this program because there is no discrete owner. Some of the properties, we find that in the address data set, the master database we have, when we go out in the field we find that the property is either inhabitable or actually vacant, [but] the structure has been removed. So besides the water cards, we do canvassing, we do physical canvassing before locations are staked.

AECOM argued that it was doing something "very unique" in Flint, which was partnering with local firms to do the work. Wong emphasized, in his presentation at a City Council meeting, that the company's desire was to hire and train Flint residents to join the lead remediation efforts: "We look for a mechanism of how we connect employees with ventures that are contractors.... In this mechanism, we look at training individuals that could, in fact, work with the existing contractors or new contractors that come onboard. So yes, that is something that's very dear to us ... linking a community with this work."[66]

Although AECOM's plan was to work with Flint contractors and residents, the firm failed to make sure that those individuals were paid for services rendered. Instead, by the time Phase 5 rolled around, the project manager for Flint-based Goyette Mechanical—Joe Parks, a city native—was forced to plead for his company to get paid, since AECOM had not been helpful in getting the city to remunerate it for work rendered. In a September 5, 2018, city council meeting, Parks mentioned that a backlog of work had developed because of equipment issues stemming from the city's change of specs for contracts that had already been negotiated:

> Last week, we were at $90,000 in accumulated extra work for not having the vac truck available to us. Since that time, we've accumulated another $17,000 in extra work for that item, so we're up to $107,000 that we intend to aggressively pursue with the city. At this point, we haven't had any communication [or] any acceptance of these charges from the administration. . . . To date, the administration has given us a solid "no, we're not gonna pay you for any of this." We would like to defer to AECOM, that's what they've been brought in for is to deal with these issues. They wrote the RFP [Request for Proposal] for the work. They wrote the specification for the project. And they wrote our contract or weighed in heavily on it. And I want to know, because no one's told us this, is whether these issues that we've brought to the table, these new constraints that have been put on us under this contract, are they legitimate changes? And if not, can you please point out in these contract documents where they can tell us not to use a vac truck, where they can tell us to dig a larger hole without compensation for it.[67]

Parks mentioned that the City of Flint issued a memo to all service line contractors on June 21 that announced the new specifications for pipeline excavations. Essentially, the city was asking contractors to expand the scope of exploratory excavations. According to Parks, the scope of excavations "was expanded out to exposing ten feet of line, which I think is a good decision by the city to do, to make sure that we're not missing any of these places. . . . But as a result of that, it's costing us significantly. In previous phases of work, the requirement's always been dig approximately two to three feet on either side of the curb box."

Due to the enlarged excavation areas and the city's refusal to pay the difference in costs, Parks noted that service line contractors were being expected to take on additional work without compensation: "We're experiencing significant costs as a result of that, to the tune of 45 grand through last Friday."

Given all the concerns that had developed since AECOM took the helm of the pipeline replacement process, residents urged the city to move on from the engineering company. They believed it had been taking advantage of the city during its water emergency. Nayyirah Shariff, of activist group Flint Rising, noted at a December 2018 city council meeting that "AECOM was supposed to do a lot more." As she put it, "As far as public relations and making sure that the public was taken along with this journey, I felt that y'all [city council members], as somebody who holds the purse strings, need to immediately terminate that contract. You need to sue them for breach of contract because this is just a classic case of what activists [call] disaster capitalism."[68] Another resident agreed:

> Everybody knew when this started, that people were going to be making money, a lot of it, in cleaning up after this misery. It is the way it works. We can call it what we want. I think disaster capitalism is a good term, but somebody's got to clean it up.... 6,000 lead service lines means 6,000 lead service lines. They had the expertise; they could have done it. If you're not gonna hold people to their contracts, then what is the point of getting proposals? Because they're just going to offer you what they think you want, and then they're going to do what they want to do. And then they're going to come back, and they're going to say, "Well, it was approximate. It was close enough. Pay us our money." That's disaster capitalism.[69]

On January 9, 2019, AECOM asked for an additional $4.8 million and a one-year contract extension,[70] promising to use Abernethy and Schwartz's predictive model going forward. But Flint's city council failed to approve this measure in a deadlocked 4–4 vote. Some council members voted no because they alleged unethical fraternizing between Mayor Weaver and firms under contract that had donated, through its employees, thousands of dollars to her.[71]

Still, due to complaints from Flint residents, city officials reconsidered its stance on the predictive model and directed AECOM to use it for water service line repairs starting with Phase 6 of the process in 2019. Without this model, Flint would be on the hook for cost overruns, like many other cities that have hired AECOM to supervise large-scale restoration projects in recent years, including Miami-Dade County,[72] Oakland,[73] Fort Worth,[74] New York, and Detroit.[75] It appeared that Flint's project wasn't the only AECOM effort to run long and over budget.

The Pipe Replacement Saga Continues

My examination of Flint's pipeline excavations shows that concerns voiced by the state regarding the costs of repairs barely scratched the surface of the city's misguided efforts. Flint's digging up of pipelines that didn't need to be repaired was only part of the city's inept attempts to replace its water infrastructure. The city also failed to prioritize the areas most impacted by lead-contaminated water during this public health disaster and neglected to go where the problems were clustered, as indicated by elevated blood lead levels. Further, under Weaver's leadership, it did not prioritize pipeline replacement efforts in areas that had been more devastated than others by the water crisis.

It makes sense that the City of Flint would assert an interest in prioritizing its pipeline repair efforts in areas with homes built before 1950. As illustrated in map 7.1, these were hot spots for elevated BLLs and water lead levels above 15 parts per billion. With the exception of a few areas on the southeast side, these hot spots also heavily coincided with areas that were primarily Black (Wards 1, 2, 5, and 6).

Despite the known demographic characteristics and structural conditions associated with lead exposure in Flint—which include proximity to water main breaks and high levels of lead in test results—these facts did not appear to influence the city's decisions regarding where line repairs took place. During Phases 1–3, lead service line replacements were reliant, reportedly, on the Abernethy and Schwartz's predictive model that prioritized age of housing. However, as seen in map 7.2, even when the city claimed to be guided by this predictive model, pipeline excavations did not necessarily correspond with the hot spots of children with elevated BLLs.

Other factors seemed more prominent with regard to the areas receiving priority attention. As a result, the city missed opportunities to target and prioritize its pipeline repair efforts on homes with lead-contaminated water and children with elevated BLLs. Instead, contractors were directed to repair homes *neighboring* these hot spot areas.

Statistical findings further demonstrate that, despite reports of progress from city leaders and AECOM, efforts to fix the service lines in *some* communities had been delayed. As we can see in table 7.1, most of the repair efforts during Phases 1–3 were located in zip codes 48506 and 48507 (in fact, residents in 48506 were twice as likely than the rest of the city receiving repairs during Phases 1–3 to receive full replacement). The highest concentration of elevated BLLs during 2014 to 2016 were, however, in zip codes 48503, 48504, and 48505.

During Phase 4, residents in the 48505 area began to receive repairs. However, they were not more likely than residents in other areas to receive full pipe replacements. Furthermore, while the percentage of residents who received repairs to the public portions of their service lines in zip codes 48503 and 48504 was higher than other areas Phase 4, pipe replacements in 48507 were primarily limited to private portions of water lines throughout Phases 1–4.

Table 7.2 shows how the population compares across space in Flint, illustrating how selected characteristics in 2018—including economic, social, and health indicators, as well as property ownership patterns—contrast by zip code. Vacancy rates, for instance, vary from 44 percent in zip code 48505 to

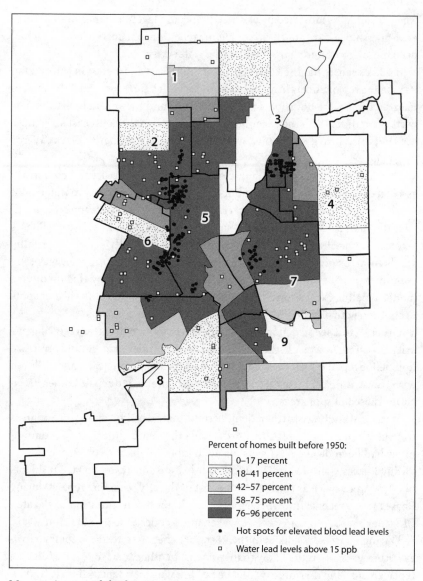

Percent of homes built before 1950:

- ☐ 0–17 percent
- ☐ 18–41 percent
- ☐ 42–57 percent
- ☐ 58–75 percent
- ■ 76–96 percent

- ● Hot spots for elevated blood lead levels
- ☐ Water lead levels above 15 ppb

Map 7.1. Areas with highest concentration of children with elevated blood lead levels and select demographic characteristics (race and age of housing stock)

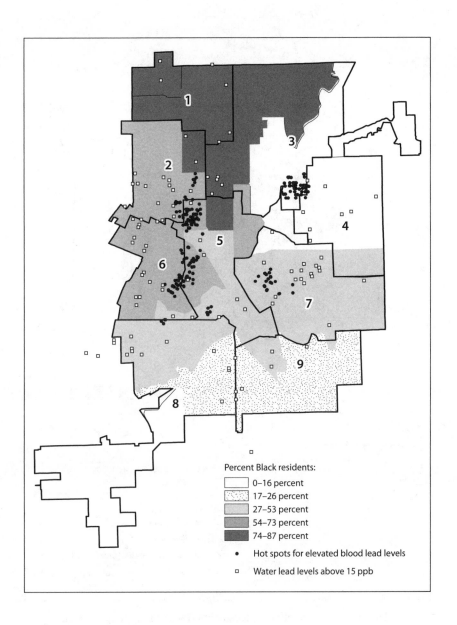

Percent Black residents:

☐ 0–16 percent

▨ 17–26 percent

▨ 27–53 percent

▨ 54–73 percent

▨ 74–87 percent

• Hot spots for elevated blood lead levels

□ Water lead levels above 15 ppb

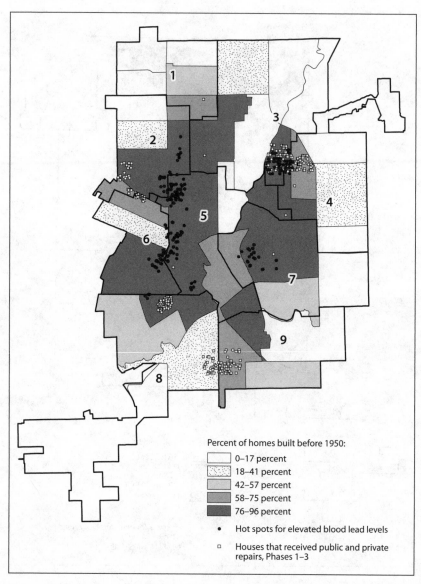

Percent of homes built before 1950:

- ☐ 0–17 percent
- ⣿ 18–41 percent
- ▨ 42–57 percent
- ▨ 58–75 percent
- ■ 76–96 percent

- ● Hot spots for elevated blood lead levels
- ☐ Houses that received public and private repairs, Phases 1–3

Map 7.2. Distribution of pipeline replacements (Phases 1–4) by age of housing structure and phase

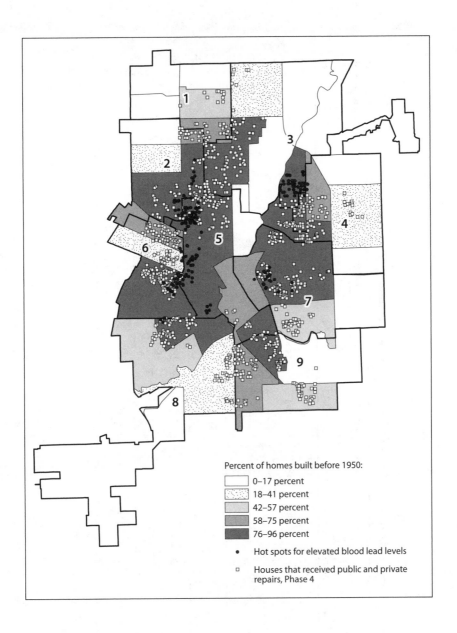

Percent of homes built before 1950:

☐ 0–17 percent
⬚ 18–41 percent
▢ 42–57 percent
▣ 58–75 percent
▰ 76–96 percent

● Hot spots for elevated blood lead levels

☐ Houses that received public and private repairs, Phase 4

Table 7.1. City of Flint service line repairs, Phases 1–4, by zip code and replacement type

	Zip Code	No. of Repairs	Full Line Replaced			Public Portion Replaced			Private Portion Replaced		
			%	OR	AOR	%	OR	AOR	%	OR	AOR
Phases 1–3	48503	160	19.0	REF	REF	17.3	REF	REF	5.3	REF	REF
	48504	262	14.6	0.33***	0.303**	35.8	1.539**	1.282	5.3	0.613	0.146
	48505	12	0.3	0.12**	0.174	0.0	—	—	0.0	—	—
	48506	302	43.1	1.39*	2.26***	32.5	0.999	0.529*	15.8	1.605	0.343
	48507	268	22.7	0.578***	0.398*	14.4	0.374***	0.474*	73.7	8.819**	10.848
Phase 4	48503	568	27.2	REF	REF	25.5	REF	REF	27.2	REF	REF
	48504	487	23.3	0.835	0.856	25.0	1.305**	1.221	23.3	0.427**	0.633
	48505	373	17.8	1.288*	1.049	14.9	0.829	0.882	17.8	0.277***	0.682
	48506	306	14.6	0.715**	0.657*	16.7	1.489***	1.54**	14.6	0.831	1.278
	48507	283	13.5	0.689**	0.77	14.8	1.355**	1.158	13.5	0.902	1.713

Source: City of Flint, FAST Start Pipe Replacement Website, https://www.cityofflint.com/gettheleadout/.

Notes: OR = odds ratio. AOR = adjusted odds ratio, adjusted for percentage of homes built before 1950 and percentage vacant

$* p < .10$ $** p < .05$ $*** p < .001$

Table 7.2. Selected characteristics of Flint residencies and residents by zip code, 2013–2015

	48503	48504	48505	48506	48507
Number of occupied units	1,193	1,411	607	887	1,586
Percentage vacant	22	28	44	33	22
Percentage employed in service occupations	23	29	38	32	20
Unemployment rate	18	19	33	23	21
Percentage of households below poverty level	40	39	41	31	29
Percentage of children under 18 below poverty level	56	60	60	44	44
Percentage of homes built before 1950	46	26	33	30	18
Percentage Black	36	73	94	8	27
Percentage white	57	21	3	76	69
Mean household income (2017 dollars)	$21,982	$23,479	$14,096	$17,472	$19,611
Median household value at zip code level	$43,500	$33,000	$22,700	$44,100	$54,200
Mean blood lead levels among children 12 and younger	2.5	2.6	2.6	2.7	2.3
Privately owned home	89.0	87.7	90.6	87.8	86.0
Home owned by Genesee County Land Bank	1.4	2.3	3.9	1.6	1.1
Home owned by a nonprofit (church or residential facility)	0.5	0.4	1.3	3.1	0.4
Home owned by a trust	4.4	1.3	2.6	2.3	2.4
Home owned by a bank or financial company	1.4	1.9	0.3	1.8	3.1
Home owned by a general business	1.4	3.5	1.0	1.5	3.3
Home owned by a rental company	1.9	2.9	0.3	1.6	3.8
Home owned by the state of Michigan	—	—	—	0.2	—

Sources: MDHHS Childhood Lead Poisoning Prevention Program; Michigan Brownfield Redevelopment Program; City of Flint Water Department.

22 percent in 48503 and 48507. In area 48505, 94 percent of the population was Black American, with a mean annual income of just under $15,000. Meanwhile, residents in zip codes 48506 and 48507—where pipe repair efforts were concentrated during Phases 1–3—were predominantly white.

A Recovery Far from Full

Back when Flint residents were walking the picket line with Dr. Karen Weaver, many believed she would be a transparent and effective leader for the city as it worked to rebound from its water crisis. Citizens had high hopes for her. Principal among these was seeing the water crisis finally resolved. But, that resolution never happened. Instead, the process of the water recovery became plagued by miscommunications, scandal, and waste during Mayor Weaver's tenure.

She subsequently lost her bid for a second term on November 5, 2019. State representative Sheldon Neeley, who had served on the Flint City Council for nine years, won by 300 votes. When the *Flint Journal* caught up with the mayor-elect, he expressed gratitude for the opportunity to serve and excitement for the chance to help Flint heal from the conflicts and chaos.

"I'm looking forward to bringing the city back together," Neeley said. "There was a lot of fracturing through this process but now we need to come together as a total united city. . . . I'm excited to work with the people in moving our communities forward."[76]

But after briefly distracting residents with allegations of fraud on the part of the Weaver administration—Neeley accused it of either misplacing or hiding nearly $20 million that should have been in the city's water fund[77]—the efforts of Flint's new mayor have hit a few snags, including the inability to discard expired bottled water before passing it out to residents.[78]

Once the City of Flint secured funds from the state and federal governments to pursue the replacement of its lead and galvanized steel pipes, residents had good reason to expect the city to get the job done quickly. The water quality issues had taken their toll; residents were ready to move on. But the repairs were delayed and delayed.

Many of the delays resulted from the administrative decisions made by the City of Flint and the State of Michigan to limit the number of homes eligible for service line replacements. Flint residents reasonably expected that the city would start at one end of a street in every zone and move down it to every service line in need of repair. Instead, at the beginning of this process, the city directed contractors to bypass homes with delinquent water bills. Flint

Figure 7.2. Flint residents waiting for (expired) bottled water.

authorized AECOM to overlook these households, despite knowing that water lines were connected to the system *and* that any unaddressed contaminated service lines could potentially contaminate neighboring ones.

Since Flint residents continued to be told to avoid tap water and depend on bottled water—while still facing high water bills—they became fed up. They felt disposable. And the attention once focused on their water issues moved on.

When the city was in the national spotlight, when President Obama was there in person drinking the water, when reporters were sitting in on city council meetings, residents thought Flint might get what it needed to overcome its water crisis. They believed Mayor Weaver's vow to "make Flint whole"[79] and AECOM's plan to leave "no ward behind."[80] But these plans came up short. As a result, when Phase 6 of Flint's FAST Start pipe replacement program began in 2019, residents testified that they had lost faith in it.

"We're back to where we first started, where we're yelling and screaming," Melissa Mays told a *New York Times* reporter. "And it seems like nobody can hear us."[81]

CONCLUSION

"Flint Ain't Fixed"

For a moment, it appeared as though help was on its way to Flint. Well-meaning folk sympathized with its residents and kept the struggle alive. Donations flooded to the city. Heart-wrenching testimonies that detailed experiences with tainted water that were posted by Flint residents on social media were shared; documentaries and news reports highlighting the most sensational aspects of this public health disaster were watched across the nation.

People who bought those "Flint Lives Matter" shirts and placards assumed that a movement that would hold officials accountable was underway. They also thought that the working-class people affected by the misguided decisions of water regulators and local administrators would finally be vindicated. Alas, this optimism was misplaced.

As soon as water testing generated results below the federal action level set by the Lead and Copper Rule of 1991, the state began its steady withdrawal of support from Flint. This included, during 2018, announcing that it would stop payments covering the water credits on active customer accounts, stop funding Flint's water being sourced from the Great Lakes Water Authority, and stop providing the city free bottled water. Finally, in December, the State of Michigan repeated its declaration that Flint's water quality had been "restored," despite persistent concerns still being voiced by residents—and even state agencies—concerning the city's pipe replacement efforts.

The challenges to these decisions from Flint residents and then-mayor Karen Weaver did not motivate any further sustained state support for the city's recovery from the water crisis. Even when dozens of angry Flint citizens traveled to Lansing—marching from the Anderson House Office Building to the steps of the capitol—to protest the state's decision to end its free bottled water program, the state still declined to resume distribution. Testing efforts have been reduced substantially since monitoring began in 2016. Yet the State of Michigan maintains, with less and less evidence, that the water crisis in Flint has passed.

But . . . Is Flint Fixed?

A few years after the water crisis initially emerged, actor Jenifer Lewis sang "Flint Ain't Fixed" in a Facebook post on July 2, 2018:

> I just got back from Flint, yeah, that Flint.
> I saw that dirty water with my own eyes.
> But not to my surprise,
> Gov. Snyder, you're not much of a provider.
> 'Cause Flint ain't fixed!
> Flint ain't fixed!
> Flint ain't fixed!
> The children up there are sick.
> The lead in the water is thick.
> Flint ain't fixed!

Over two years after Ms. Lewis first shared that song, Flint residents would agree—it still "ain't fixed." Flint residents continue to be concerned about how water quality during this period in the city's journey has impacted their health and how it will shape their children's futures. Recent reports indicate that reading and math proficiency among Flint's third graders has dropped by half, to 10 percent, surpassing statewide declines between 2013 and 2017.[1] Despite these trends, the Flint Community Schools system lacks the resources to provide sustained support for its students in the years to come. With declining enrollments and financial deficits,[2] the district is unlikely to have the money or staff to help its kids maintain adequate academic performance, much less to address the special education needs that have emerged due to the lead crisis.

Flint residents, by early January 2020, continued to rely on botted or filtered water every day to make their coffee or tea, to bathe, to brush their teeth, to prepare and cook their food, and to feed their pets and water their plants. Appliances like water heaters, for some residents, were ruined. Home values plummeted and are not likely to improve anytime soon. This is their reality—in large part—because their drinking water was contaminated by the (in)actions of the state, by its coordinating agencies, and by the slow pace of the city's recovery.

Due to these circumstances, many Flint residents continue to pick up water from church-sponsored bottled water giveaways and show up to city council meetings to voice their concerns. In the absence of bottled water giveaways, residents still suspicious of the state's declaration about the water quality rely on store-bought water or professionally installed filters to process their drinking water. These are just a few of the ways they have managed their anxieties about the potential—and unaddressed—dangers that lurk in the tainted taps of Flint.[3]

It Ain't Just Flint: Lead Pipes and Drinking
Water Emergencies in the United States

As a result of efforts by the lead industry—most notably the Lead Industries Association, which worked tirelessly for decades to undermine attempts to regulate the use of lead by manipulating plumbing codes and training regulations—Flint is not the only city that ought to be concerned about the issue.[4] Although by the late 1890s it was known that water pipes were the leading cause of lead poisoning, most municipalities of at least 30,000 people still enforced plumbing codes requiring the use of lead in the construction and repair of water distribution systems. By the 1920s, cities were finally taking direct action, revising plumbing codes to control and prohibit the use of lead.[5] Nonetheless, most cities also opted to overlook the *known* public health risks associated with leaded pipes and materials already in use. This happened because of the Lead Industries Association's capacity to convince most housing authorities to adopt plumbing codes that required the use of lead in large developments and to challenge restrictions on lead products.[6]

A national audit conducted in 1986 by the US Environmental Protection Agency revealed that 73 percent of the 153 municipalities studied—representing forty-one states, the District of Columbia, and Puerto Rico—still utilized lead service lines known to cause problems even with relatively noncorrosive water.[7] This study collected data through 1984, just two years before the installation of lead pipes nationwide was banned by the Safe Drinking Water Act Amendments of 1986.[8]

While some municipalities, especially cities in the South, claimed that lead service lines no longer remained in service (with Tallahassee, Meridian, Charleston, Memphis, and Richmond being exceptions), the vast majority of surveyed municipalities acknowledged that a significant percentage of lead service lines had never been replaced with copper or PVC material. Additionally, lead solder, even in the construction of new homes, had not substantially declined either.[9]

Due in part to these and other realities, Flint is not alone in its uphill fight against childhood lead poisoning. Among preschool children, lead exposure continues to be a critical environmental hazard, with the danger often concentrated in the poorest and oldest areas of municipalities. For instance, in Georgia in 2014, health department officials reported that over 2,500 children had lead in their system, with most high-risk areas concentrated in the metro Atlanta area.[10] Meanwhile, in Indiana, clusters of childhood lead poisoning have been found in Gary, where in July 2019 it was reported that almost 25 percent of children had elevated blood lead levels, a situation that promoted federal funding for the remediation of lead exposure in the city.[11]

These examples indicate that the two highest-profile efforts to curtail lead in the environment did not complete the job of protecting children from this danger: laws that forced paint manufacturers to remove lead from paint, along with requiring home renovators to use extreme measures when dealing with surfaces that had been covered with lead paint, didn't completely solve the problem of childhood lead poisoning in America; neither did the banning of lead gasoline in 1996.

Lead is insidious. Even parents working in factories with lead products could contaminate their children after leaving the job with the metal on their clothing. In one specific case, Water Gremlin LLC—which had previously been hit with a $7 million fine by the State of Minnesota for its seventeen-year practice of dumping carcinogenic industrial solvents—was ordered to shut down in October 2019 after twelve children of its employees were exposed to lead. The company manufactured products, including fish sinkers and battery terminals, using the solvent trichloroethylene.[12]

Children also face increased risk of being exposed to lead if they attend schools with lead-contaminated drinking fountains. Most states have struggled to rid school facilities of lead because, like Connecticut for example, they do not require schools to have their water systems checked.[13] Meanwhile, some districts have begun to act on their own. Multiple districts in Florida have been tested since 2016 after elevated levels were found in some schools' drinking fountains. However, there have been minimal efforts by the state to ensure schools install filters, and Florida law doesn't require schools to test drinking water for lead.[14] Rural states, such as Montana—where school water tests happen on a purely voluntary basis and many smaller, rural schools do so rarely, if ever[15]—have also been heavily criticized for not addressing lead in public schools.[16]

Beyond concerns about the presence of lead on surfaces in buildings and school fountains, fears continue to be fueled by cities that have avoided mitigating sources of lead exposure in high-risk areas. Many municipalities have been gaming the system, persistently refusing to perform Lead and Copper Rule compliance testing correctly. Flint certainly is not only the city to get caught cheating on this important preventive measure.

In 2017, an audit of 105 water departments in Georgia discovered that 58 of them systematically tested low-risk sites instead of high-risk ones. Additionally, 49 of those water systems sought to cover up their efforts by labeling the lower-risk sites as high-risk.[17] Likewise, in Connecticut it was found that many water systems had violated lead-level requirements at least once since 2013. A total of 14 water systems in the state were not in compliance with the Lead and Copper Rule.[18]

In Pennsylvania, water problems began developing when the Pittsburgh Water and Sewer Authority damaged water pipes in 2014 by altering the chemicals used in its system without realizing the repercussions. Pittsburgh did not publicly acknowledge the water problem—until the city made plans to distribute water filters to residents in 2017 in advance of the bottled water distribution that began in early 2018. The Natural Resources Defense Council filed a lawsuit against the city; the settlement in February 2019 required Pittsburg to replace affected lead service pipes, prioritize homes with children, and pass out free water filters to low-income residents.[19]

Finally, in the same way that Flint is not the only city in the United States that failed to remove existing lead pipes (despite the well-documented public health risks associated with the material) or the only city to mismanage its obligations by failing to address the sources of lead contamination in high-risk areas, Flint is not the first majority Black and poor city forced to prove that service providers were selling lead-contaminated water. Nor is Flint's water crisis the first time local-level officials and the EPA have neglected to protect citizens from lead-contaminated water.

In 2003, a decade before Flint residents began getting sick, Washington, DC, residents were voicing discontent with high lead levels in their drinking water. The *Washington Post* revealed that while the District of Columbia Water and Sewer Authority (DC Water) stubbornly assured customers over several years with quality reports that stated the water was safe, the fact was that the utility was aware of potential lead contamination. Freedom of Information Act documents secured by the *Post* indicated that, as in Flint, DC Water would drop homes that had previously tested high for lead so as to avoid testing these high-risk homes, which changed compliance results in order to avoid water quality violations.

DC Water initially learned about water quality concerns when the city tested 53 homes in 2002 and found that over half had water lead levels exceeding the federal trigger level of 15 parts per billion. When DC Water tested for lead again in 2003, the utility found that two-thirds of 6,118 homes had elevated water lead levels, requiring public disclosure and lead remediation.[20] Dr. Marc Edwards and his colleagues blamed this crisis primarily on a November 2000 switch in disinfectant from free chlorine to chloramine, which changed the water chemistry and caused lead leakage from service line pipes.[21]

Substantial public health interventions did not emerge until the story became front-page news in early 2004, months *after* the *Washington Post* article.[22] Officials issued warnings to pregnant women, nursing mothers, and children under six years old, and DC Water distributed over 30,000 free water filters to customers and tested fountains and sinks in the city's schools. Eventually phosphate was added to the water and sections of municipal pipes containing lead were gradually replaced. In recent years, DC Water has developed a voluntary service pipe replacement program to help homeowners get rid of privately owned lead pipes.[23]

As in Flint, an unconventional governance style likely contributed to Washington's problems. Since the EPA, instead of a state agency, regulates DC Water out of its Philadelphia office, scholars believe weak oversight and communication between the city and the EPA contributed to the crisis. Ultimately, water system managers failed the citizens of both Flint and the nation's capital by neglecting to implement corrective measures. Additionally, the regulatory environments in these municipalities lacked transparency and accountability, as evidenced by EPA records documenting how the State of Michigan discontinued its obligation to inform citizens of lead exposure violations due to lack of funding.

These conditions make it difficult for any community group—with or without collective efficacy—to fight for social justice. Although Flint and DC community groups raised awareness about their right to clean, affordable water, their capacity to motivate change was undermined by their lax regulatory culture. Movement didn't occur until the depth of the lead problem was exposed fully for all to see. How do these patterns of environmental injustice emerge?

Many scholars in various disciplines have documented the extent to which people of color and poor communities face more environmental dangers than similarly situated white ethnic communities. Due to decades of residential and occupational segregation, minority and low-income neighborhoods often occupy the "wrong side of the tracks," receiving different treatment when it comes to enforcement of environmental regulations. Political and socioeconomic power play instrumental roles in shaping the distribution of residential amenities and harmful infrastructure across communities. Studies have shown that low-income people and communities of color bear the disproportionate burden not only of general air pollution, lead poisoning, pesticide poisoning, and groundwater and soil contamination but also of hazardous waste disposal and garbage incineration placements. As a result, reproductive disorders, respiratory illnesses, cancer, asthma, and leukemia are prevalent and concentrated in low-income communities of color. And yet these groups have been only marginally involved in the nation's environmental movement. The day-to-day lives of poor Black Americans, for example, differ significantly from those of the well-publicized environmentalists, who champion such issues as wilderness and wildlife preservation, resource conservation, world population control, and industrial abatement. Many of the battles waged by mainstream environmental organizations during the height of the environmental movement in the early 1970s had marginal effects on the deteriorating conditions in urban areas. They were issues that had very little to do with the plight of people of color or working-class communities. Meanwhile, spaces where the poor and people of color reside, like Flint, are evermore stigmatized due to persistent neighborhood decline caused by uneven development and decades of benign neglect. By the 1980s, Flint was beginning to show the effects of this vicious cycle. Certain areas of the city were marked by graffiti, uncollected trash, and overgrown grass and weeds. Unkept properties attracted thieves and squatters. White and middle-class flight certainly contributed to vacancy rates, but spaces like these—nested within discarded cities like Flint—also struggled because they were targeted for disinvestment.

Instead of supplying the city with assistance to address its long-term issues, the State of Michigan drastically reduced its funding, a move that exacerbated Flint's financial problems and made the city subject to the emergency manager law. The state then used this legal construct to impose its agenda on Flint, that included adaptations to the city's water service protocols as well as attempts to circumvent water protection policies and procedures. This was done by side-lining the regulatory community and using local water experts to mislead the public before—and throughout—the city's water recovery process.

It is important to note that, while the actions of the state-appointed administrators in Flint played a significant role in ruining its already deteriorating water system, their presence and actions alone did not cause water costs to spike and quality to not just decline but remain unsuitable for human consumption. Flint, stymied by a shrinking population and tax base, declined after years of cost-cutting, infrastructure neglect and, additionally, programmatic disinvestment in public improvement projects and essential services. Furthermore, the state's actions were indirectly encouraged by the EPA's history of lax regulatory controls coupled with bureaucratic obstruction by regulatory administrators, who mismanaged Flint's water system and undermined efforts by water justice advocates who were seeking an equitable and sustainable solution to the city's issues.

While opposition and inaction to water justice might be informed by perceptions of concerned residents and the nature of social cohesion within community groups, this book demonstrates that the most influential pushback to justice advocacy came from Michigan's political environment—one shaped by the actions of an interagency elite that prioritizes municipal budgets over public safety and governs by coercion and shrewd manipulation. Evidence in this study illustrates that the rewards of water justice advocacy were undermined by false assurances and governing decisions designed to manipulate residents into setting aside their concerns. Notwithstanding the steps made to address residents' verified concerns, actors within the intergovernmental regime continued to downplay water system deficiencies, deflect responsibility for fixing problems, and delay the deployment of necessary resources for a poverty-stricken community (while declaring the water safe, in spite of the many repairs that still needed to be made). Making matters worse, after agreeing to only fix a fraction of the service lines that required it—and failing to offer free bottled water until the replacement process is complete—the State of Michigan compounded the poor conditions wrought by the crisis by condoning Flint's initial decision to restrict service line repairs to homes with current water accounts.

Finally, it is important that we take full stock of Michigan's public disagreement with Flint's dig practices during Phase 5. Based on complaints voiced by state parties, its primary concerns centered around the cost efficiency of the city's contractor payments and not the city's consistent inability to concentrate its efforts in spaces that were contending with high rates of childhood lead poisoning and lead pipe hazards. In addition, the state did not express fault with decisions made to deprioritize repairs in the city's poorest and oldest areas during the initial phases of the process. This decision was likely cost-effective from the state's perspective since officials did not have the capacity to draw funds from the Children's Health Insurance Program earmarked for Medicaid-eligible homes until Phase 4. However, this decision, in concert with efforts to exclude properties without active water accounts, appears especially biased against Flint's poorest and most disenfranchised residents (as well as being a shortsighted and life-altering lead remediation policy).

Most residents in discarded cities are under enormous pressure to find reasonable community amenities and family-sustaining jobs. They aren't set up to withstand a long-term public health disaster. In fact, in Flint, people could barely withstand conditions *before* fears about the water supply emerged. People had been leaving the city in droves; it was increasingly hard to make ends meet due to cuts in municipal and transportation services as well as shortages of good-paying jobs. The water crisis made circumstances so much worse. Foreclosure rates had started to stabilize before the water source switch,[24] but a steep 161 percent increase in real-estate-owned properties between 2017 and 2018 indicates that this trend reversed course after the water crisis.[25]

Since the Flint water crisis has faded from the national spotlight, other cities throughout the United States have discovered lead issues within their water systems. In August 2019, officials in Newark, New Jersey, had to resort to passing out bottled water because residents were being affected by lead. The confusion began in the fall of 2018 when the City of Newark began distributing water filters, although they had not given a reason to residents as to why they were receiving the water filters—or how important it was that they use them.

It wasn't until Newark residents received a letter in August 2019 from the EPA that they learned their water could make them seriously ill and that the water filters the city had been passing out did not effectively protect them from the contaminated water. Even in homes that had correctly used the filters, water tests produced results above the federal limit for lead.

As in Flint, Newark residents have had to wait—occasionally for hours—to get bottled water provided by the city. To help residents get the assistance and resources they needed, the Natural Resources Defense Council filed a federal lawsuit to compel the city to distribute bottled water evenly and to not avoid, as it had previously, distributing it in the eastern sections of Newark.[26] Even with outside help from the Natural Resources Defense Council, residents have had to encounter another major hiccup, with expired bottled water curiously getting shipped and handed out to residents in need.

If Not Now, When? Social Justice for Flint?

Deceased rapper—and Flint native—MC Breed couldn't have been more right when he boasted of his city's tenacity, pride, and grit in his 1991 hit song "Ain't No Future in Your Frontin'":

> Suckers causin' static, cause they still be disagreein'
> I don't give a —— cause I'm from F-l-i-n-t' n
> A city where pity runs low
> If you ever shoot through my city, now you know
> Cause we are strictly business and we also got our pride
> And if you don't like it, I suggest you break wide . . .
> That's the way I am, MC Breed cannot be different
> Never change my ways for the world or the government
> If I was the president, then I would state facts
> You leave it up to me, I paint the White House black.
> It ain't no future in your frontin'.

Compared to more affluent communities, residents in poor working-class cities like Flint tend to lack resources such as insurance, savings, and access to social networks that can help them manage during an emergency.[27] And yet, they are surviving another round in what appears to be an endless and peculiar game of exploitation and immiseration only because of the individual hustle and collective unity that emerge when a community's voice needs to be heard. But a city like Flint can only be so strong.

Flint's citizens continue their fight for water justice—only a part of their persistent struggles against disinvestment and abandonment—but when will they finally receive the support and access to the hard, factual information they deserve to address the fundamental factors behind these issues? When will the state and the city, as MC Breed would say, "stop frontin" and provide the essential services residents need? When will the state and city implement measures to encourage sustainable and equitable growth in this poor community? When will residents be effectively included in the governing process with the consistent presence of a city ombudsperson, a functioning code enforcement department, and solution-oriented measures like participatory mapping[28] and expanding residents' access to local public health data, which will help encourage trust and ensure accountability?

In understanding that the effects of the water crisis were concentrated in disempowered and disinvested spaces, it is unacceptable to ignore the need to both expose and counter the long-term consequences of benign neglect and uneven development. Due to unresolved residential clustering in Flint—which is based on race and class conflicts—racial minorities and low-income residents were disproportionately exposed to lead during this public health event. Similar to the Ninth Ward residents in New Orleans, who faced a greater risk of being flooded than higher-lying, more affluent areas of the city during the Katrina disaster,[29] preexisting differences played a substantial role in determining which Flint communities would be disproportionately exposed to lead during the water crisis. The hard fact is, in America's cities, a public health or natural disaster tends to magnify and reinforce preexisting inequalities along racial and class lines.

Just like Flint is not the only American city to endure water quality issues, it is not the first American city to allow its inequalities before a disaster influence how disaster remediation plays out.[30] For example, researchers found that during the Katrina ordeal, Black Americans were more likely than whites to reside in flood-damaged areas and therefore experienced more significant hardships in attempting to rebuild after the storm.[31] Likewise, after Hurricane Andrew, researchers revealed that white citizens in Florida and Louisiana were far more likely to receive insurance settlements than were non-Cuban Latinos and Black Americans, who generally had to turn to support from interfaith churches, nonprofits, and their families to get through the disaster. Because of their lack of transportation and work conflicts due to childcare issues, poor residents even had difficulties gaining access to disaster assistance centers.[32]

In the United States, we often assume that meritocracy and community cohesion will help citizens resolve neighborhood problems. But the research included in *Tainted Tap* illustrates that self-help and community efficacy can heighten awareness, but they cannot eradicate the tangled legacies of racism, class bias, and benign neglect in the discarded cities where regulations and procedures undermining "market freedoms" have long been implemented.

The stress and setbacks created by the response to the Flint water crisis by the administration of Governor Rick Snyder revealed a biopolitical agenda, a logic of disposability, and a bureaucracy's tendency to self-protect—even at the expense of vulnerable citizens. Instead of safeguarding disadvantaged residents, the state all but abandoned them.

Poverty has become a convenient excuse for bureaucratic mismanagement in discarded cities. Little attention is paid to the pro-business, anti-regulation regime that triggers public health crises, not to mention the way the state's method of encouraging economic growth and financial efficiency weakens essential services in poor communities of color. It is undeniably clear that local- and state-level administrators played a crucial role in creating Flint's human-designed water crisis. It is also clear that efforts were made by those acting on the state's behalf to undermine community efforts to bring attention—and help—to the city.

Essentially, the State of Michigan—through its appointed agents, rubber-stamping regulatory agencies, and disempowered local water treatment professionals—was the chief architect of the water source switch. It was understood that Flint's water treatment plant could not successfully process water safe for human consumption from the Flint River. And still the state wielded its power over local authorities and regulatory agencies to force the water source transition. Afterward, state and local governments—collectively and repeatedly—assured Flint residents that their water was safe to drink and deflected their responsibility to address water quality issues onto homeowners.

State-level administrators, particularly within Michigan's Department of Environmental Quality (known as the Department of Environment, Great Lakes and Energy since February 2019), contributed to the willful neglect of Flint's families. The state authorized the city to process river water rife with life-threatening contaminants through a water treatment plant that would have required significant upgrades in order to provide safe drinking water. It ignored subsequent complaints about the water quality and belittled attempts to refute government assertions.

DEQ administrators disregarded water treatment plant concerns from water plant workers and ignored the EPA's urging for the agency to require Flint to implement a corrosion control program. Officials within the governor's office were instrumental in undermining collective efforts to address water quality concerns. Many policy shifts authorized by emergency managers required state treasury approval, meaning that when city emergency managers approved a water source shift and drastic rate increases, Michigan state officials endorsed those decisions.

Flint's city charter requires residents without approved springs or wells to purchase public water,[33] yet evidence shows that substantial state-level shifts in water safety and regulation procedures drastically diminished the quality of public water available to residents. Evidence also illustrates that efforts to repair known hazards in the system were sluggish, misguided, and biased. This restructuring, which curtailed access to safe, affordable water, represents an escalation of previous iterations of biopolitics that contributed to the devaluation, dehumanization, and loss of impoverished Black lives across the United States throughout its history. It's an episode that joins that long, insidious roll call that runs from unchecked lynchings and white vigilantism to Black women's "Mississippi appendectomies."

Despite persistent, organized community opposition, the benign neglect of vital public resources like drinking water is part and parcel of new forms of social discipline and control that encourage population decline and despair. The architects of these new forms rely on the absence of regulatory controls and on the ignorance of the subjects of this control. Health conditions caused by disparate exposure to environmental toxins in Flint and other communities with long, contentious histories of inequality and oppression must be viewed as products of the state's reconstituted power to grant life to—and potentially impose death on—subjects of neoliberal municipal regimes.

Although the Flint water crisis no longer appears in national headlines, the biopolitical hurdles that Flint residents face have not been resolved. The Flint water crisis—a manifestation of malign neglect—reverberates in the lives of its citizens in profound ways. To be sure, residents are said to be surrounded by partnerships consisting of area universities, the City of Flint, and local philanthropic foundations all working to implement programs to reduce the damage caused by lead contamination and improve the quality of health among those impacted. Reportedly, individuals and organizations have come together to address the short- and long-term needs of Flint residents, especially children, who can now benefit from an expansion of interventions like "full-service schools," designed to assist children facing multiple developmental and learning challenges.[34]

Additionally, according to Hurley Medical Center physician Mona Hanna-Attisha, "We have been able to establish free, year-round childcare. We are the only city to have universal preschool. We have Medicaid expansion, we have mobile grocery stores, and we have breastfeeding services, mental health support, and positive parenting programs." Therefore, with Flint residents and local organizations "rolling up their sleeves to make sure that our kids turn out OK" and Michigan's new governor, Democrat Gretchen Whitmer, pledging to assist Flint with this crisis, many claim to be optimistic about the city's future.[35]

But others wonder whether Flint can really be fixed.[36] Following decades of white supremacist housing practices and benign neglect, Flint's working-class and Black populations—densely situated in disinvested neighborhoods, where the worst of the water crisis was concentrated—seemed slated for removal by a political culture embracing an economic model that encourages competition and cost cutting. Therefore, in a state that maintains its right to appoint an emergency manager to initiate actions like Flint's misguided water source switch, the future is truly uncertain. Anything can happen. Only time and targeted investment will determine whether Flint can rebound from the water crisis and move beyond the other problems facing it. Godspeed.

ACKNOWLEDGMENTS

It feels good to finish a book. I am especially grateful to be at the end of this journey.

I want to begin by expressing my gratitude for the blessings that have been bestowed upon me. I am thankful for the ground that was cultivated in Flint and for the people and surroundings that shaped how I see the world; I gain strength and purpose from them. I appreciate the opportunity to write about the social problems and conditions that are important to me.

This book took much time to research and write. I cannot express my gratitude enough to the Flint residents who shared their experiences and detailed the hurdles they faced as a result of these circumstances. A huge thanks goes out to the folks who let me sleep on their couch and in their guest bedroom while I conducted interviews and went through the archives in Flint. I am also grateful for the cooperative water regulation and health administrators as well as for the hardworking archival librarians who helped me secure the data I needed to write this book. Accordingly, my gratitude extends to the editors, reviewers, and staff at the University of North Carolina Press for their encouragement and support throughout the writing process.

While researching and writing this book, I also benefited from strong institutional and intellectual support from colleagues at the University of Vermont and Florida State University.

Indeed, this effort would have been difficult to pull off without consistent help from students, colleagues, and friends. Undergraduate and graduate students helped me transcribe hours of Flint City Council meetings. They also helped me gather and process reams of studies on the health effects of neighborhood blight and secondary research concerning Flint.

Colleagues graciously offered their time to read my work and made the effort to provide constructive feedback; they listened to my complaints about delays and encouraged me to push forward. My friends urged me to make it plain and to maintain focus. And family members, like Angie McClendon Mayes, kept snapping pictures and keeping me posted on the latest in Flint.

I couldn't have completed this project without the support of my children, now teenagers, Kaitlyn and Elijah. Thanks to both of you for tolerating my commitment to this project and the long hours I spent working when I could have been spending more time with you. I appreciate the space to create and the offering of consistent love and support, despite how my work tasks and habits tended to flood our time in recent years.

Finally, to my loving and supportive husband, Taj: my deepest gratitude. It was a great relief to be able to count on you when I remained perpetually glued to my laptop. Your encouragement and support throughout this journey are appreciated. Much respect.

NOTES

ABBREVIATIONS

AJPH	*American Journal of Public Health*
AP	Associated Press
BHL	Bentley Historical Library, University of Michigan
CCE	S. Roy and M. Edwards, "Chronological Compilation of E-mails from MDEQ Freedom of Information Act Request 6526–15 and 6525–15 (2015)," Flint Water Study, http://flintwaterstudy.org/wp-content/uploads/2015/10/MDEQ-USEPA-Final.pdf, accessed August 9, 2018
DFP	*Detroit Free Press*
DN	*Detroit News*
FCC	Flint City Council
FJ	*Flint Journal*
GHCC	Genesee Historical Collections Center, University of Michigan–Flint
MDEQ	Michigan Department of Environmental Quality
MDHHS	Michigan Department of Health and Human Services
OBP	Olive Beasley Papers, Genesee Historical Collections Center, University of Michigan–Flint
SFPC	St. Francis Prayer Center Records, 1993–2001, Bentley Historical Library, University of Michigan
USEPA FOIA	United States Environmental Protection Agency Freedom of Information Request, EPA-R5–2016–003714

1. INTRODUCTION

1. Sarah Schuch, "Swartz Creek Sees Largest Amount of Snow with 18 Inches, Followed by Flint with 17.1 Inches," *FJ*, January 6, 2014, https://www.mlive.com/news/flint/2014/01/swartz_creek_sees_largest_amou.html#incart_river_default; National Weather Service, "January 4–5, 2014 Winter Storm," National Weather Service, accessed March 18, 2018, https://www.weather.gov/dtx/snowfalltotalsfromjan4–5.

2. AP, "Governor: State Stepping Up, People Should Stay In," Michigan Radio, January 5, 2014, https://www.michiganradio.org/post/flint-area-gets-top-10-record-breaking-snowfall-detroit-not-so-much.

3. FCC, transcript of public comment period, January 13, 2014.

4. FCC, transcript of public comment period, January 13, 2014.

5. Kristin Longley, "Storm Damage Leaves Flint Residents Trapped at Dead End; Some Blame Layoffs for Delay," *FJ*, July 5, 2012, https://www.mlive.com/news/flint/2012/07/storm_damage_leaves_flint_resi.html.

6. Longley, "Storm Damage Leaves Flint Residents Trapped at Dead End."

7. Kayla Habermehl, "Tight Budget Makes It Difficult to Deal with Dead Trees in Swartz Creek," *FJ*, July 9, 2011, https://www.mlive.com/news/flint/2011/07/tight_budget_makes_it_difficul.html.

8. S. Weiss, "The Flint, Michigan Water Crisis: Concurrent Private-Public Funds as the Most Effective Legal Tool to Compensate Victims," *University of Pennsylvania Journal of Business Law* 19, no. 4 (2017): 1035.

9. FCC, transcript of public comment period, January 13, 2014.

10. FOIA request for DEQ, FOIA 6797 Item 3, January 2016.

11. FOIA request for DEQ, FOIA 6797 Item 3, January 2016.

12. Mitch Smith, "A Water Dilemma in Michigan: Cloudy or Costly?," *New York Times*, March 24, 2015, https://www.nytimes.com/2015/03/25/us/a-water-dilemma-in-michigan-cheaper-or-clearer.html.

13. Flint Democracy League et al., *Petition Alleging Violations of the Human Rights of Citizens of Flint, Michigan, by the United States of America, with a Request for an Investigation, Hearing on the Merits and Precautionary Measures*, submitted by Claire R. McClinton, Jeanne M. Woods, and John C. Philo on Behalf of Bishop Bernadel L. Jefferson et al. (2017), 12–13, accessed November 15, 2019, http://admin.loyno.edu/webteam/userfiles/file/FINAL%20FLINT%20PETITION%20AND%20MEMORANDUM%2011_28_17%20with%20correct%20TOC%20(1).pdf.

14. Flint Democracy League et al., 14, 67.

15. Flint Democracy League et al., 18.

16. Flint Democracy League et al., 18, 21, 67.

17. Kris Maher, "Flint's Water Woes Make Residents Feel Like 'the Walking Dead'," *WSJ*, January 21, 2016, https://www.wsj.com/articles/water-contamination-crisis-strains-flint-residents-resources-1453401814.

18. Mona Hanna-Attisha, *What the Eyes Don't See: A Story of Crisis, Resistance, and Hope in an American City* (New York: One World, 2018); Anna Clark, *The Poisoned City: Flint's Water and the American Urban Tragedy* (New York: Metropolitan Books, 2018).

19. Ashley E. Nickels, *Power, Participation, and Protest in Flint, Michigan: Unpacking the Policy Paradox of Municipal Takeovers* (Philadelphia: Temple University Press, 2019); Benjamin J. Pauli, *Flint Fights Back: Environmental Justice and Democracy in the Flint Water Crisis* (Cambridge, MA: MIT Press, 2019).

20. Zahra Ahmad, "Planned Grocery Store Would Address 'Food Desert' in Flint," *FJ*, March 26, 2019, https://www.mlive.com/news/flint/2019/03/planned-grocery-store-would-address-food-desert-in-flint.html.

21. St. Clair Drake and Horace Roscoe Cayton, *Black Metropolis: A Study of Negro Life in a Northern City* (New York: Harcourt, Brace and World, 1970).

22. Wendell E. Pritchett, "The 'Public Menace' of Blight: Urban Renewal and the Private Uses of Eminent Domain," *Yale Law and Policy Review* 21, no. 1 (2003): 1–52.

23. Arnold R. Hirsch, "Searching for a 'Sound Negro Policy': A Racial Agenda for the Housing Acts of 1949 and 1954," *Housing Policy Debate* 11, no. 2 (2000): 393–441.

24. Dorceta E. Taylor, *Toxic Communities: Environmental Racism, Industrial Pollution, and Residential Mobility* (New York: New York University Press, 2014), 228–61; Robert C. Weaver, "Class, Race and Urban Renewal," *Land Economics* 36, no. 3 (1960): 235–51.

25. Richard L. Marcus, "Benign Neglect Reconsidered," *University of Pennsylvania Law Review* 148, no. 6 (2000): 2009–43; Deborah Wallace and Rodrick Wallace, *A Plague on Your Houses: How New York Was Burned Down and Public Health Crumbled* (London: Verso, 2001).

26. Robert B. Reich, *The Work of Nations: Preparing Ourselves for 21st Century Capitalism* (New York: Vintage Books, 1992).

27. John D. Kasarda, *Industrial Restructuring and the Changing Location of Jobs* (New York: Russell Sage Foundation, 1995).

28. Gary J. Miller, *Cities by Contract: The Politics of Municipal Incorporation* (Boston: MIT Press, 1981).

29. David Harvey, *A Brief History of Neoliberalism* (London: Oxford University Press, 2005).

30. Andrew R. Highsmith, *Demolition Means Progress: Flint, Michigan, and the Fate of the American Metropolis* (Chicago: University of Chicago Press, 2015), 11.

31. Dominic Adams, "Here's How Flint Went from Boom Town to Nation's Highest Poverty Rate," *FJ*, September 21, 2017, https://www.mlive.com/news/flint/2017/09/heres_how_flint_went_from_boom.html.

32. Lolita Brayman, "Why Haven't Flint Residents Fled?," *Washington Post*, February 22, 2016, https://www.washingtonpost.com/news/monkey-cage/wp/2016/02/22/why-havent-flint-residents-fled/; Julie Mack, "Flint Is Nation's Poorest City, Based on Latest Census Data," *FJ*, September 19, 2017, https://www.mlive.com/news/2017/09/flint_is_nations_poorest_city.html.

33. Mona Hanna-Attisha et al., "Elevated Blood Lead Levels in Children Associated with the Flint Drinking Water Crisis: A Spatial Analysis of Risk and Public Health Response," *AJPH* 106, no. 2 (February 2016): 283–90.

34. Dominic Adams, "Half of Closed Flint Schools over Last 10 Years in Predominantly Black Neighborhoods in Northwest Quadrant," *FJ*, May 12, 2013, http://www.mlive.com/news/flint/index.ssf/2013/05/neighborhoods_around_closed_fl.html.

35. Dominic Adams, "Flint Northern Students Stunned with Decision to Close School," *FJ*, March 14, 2013, http://www.mlive.com/news/flint/index.ssf/2013/03/flint_northern_students_stunne.html; Dominic Adams, "Flint Schools Sitting on Two Dozen Closed Buildings in the City," *FJ*, October 2, 2015, http://www.mlive.com/news/flint/index.ssf/2015/10/a_handful_of_closed_flint_scho.html.

36. Eric Dresden, "Closing of Second Kroger in Flint Shocks Community on Heels of Meijer Closure," *FJ*, March 4, 2015, http://www.mlive.com/news/flint/index.ssf/2015/03/flint_mayor_left_in_total_shoc.html.

37. Interview with Flint resident, May 18, 2016.

38. John Rummel, "Detroit Needs Emergency Action, Not an Emergency Manager,"

People's World, April 24, 2013, https://www.peoplesworld.org/article/detroit-needs
-emergency-action-not-an-emergency-manager/.

39. Elizabeth Kneebone, "The Growth and Spread of Concentrated Poverty, 2000
   to 2008–2012," Brookings Institute, July 31, 2014, https://www.brookings.edu
   /interactives/the-growth-and-spread-of-concentrated-poverty-2000-to-2008–2012
   /#/M10420; Myron Orfield, *American Metropolitics* (Washington, DC: Brookings,
   2002). A. Schafran, "Origins of an Urban Crisis: The Restructuring of the San
   Francisco Bay Area and the Geography of Foreclosure," *International Journal of
   Urban and Regional Research* 37, no. 2 (2013): 663–88; A. Schafran, "Discourse and
   Dystopia, American Style: The Rise of 'Slumburbia' in a Time of Crisis," *City* 17,
   no. 2 (2013): 130–48; Chris Niedt, "Legal Geographies—The Politics of Eminent
   Domain: From False Choices to Community Benefits," *Urban Geography* 34, no. 8
   (2013): 1047–69. M. Johnson, *The Second Gold Rush: Oakland and the East Bay in
   World War II* (Berkeley: University of California Press, 1993).

40. Myron Orfield and T. Luce, "America's Racially Diverse Suburbs: Opportunities
   and Challenges," Institute on Metropolitan Opportunity, Minneapolis, July 2012,
   https://www.law.umn.edu/uploads/e0/65/e065d82a1c1daobfef7d86172ec5391e
   /Diverse_Suburbs_FINAL.pdf; Chris Niedt, "Politics in the Diversifying Suburbs," in *Atlas of the 2012 Elections*, ed. J. C. Archer, F. Davidson, E. H. Fouberg,
   K. C. Martis, R. L. Morrill, F. M. Shelley, R. H. Watrel and G. R. Webster (Lanham,
   MD: Rowman and Littlefield, 2014), 120–26.

41. Becky Nicolaides, *My Blue Heaven: Life and Politics in the Working-Class Suburbs of
   Los Angeles, 1920-1965* (Chicago: University of Chicago Press, 2002); Chris Niedt,
   "Gentrification and the Grassroots: Popular Support in the Revanchist Suburb,"
   Journal of Urban Affairs 28, no. 2 (2006): 99–120.

42. Shannon Murphy, "State Revenue Sharing Cuts Reshape City Governments in
   Michigan," *FJ*, October 16, 2011, https://www.mlive.com/news/2011/10/state
   _revenue_sharing_cuts_res.html.

43. Murphy, "State Revenue Sharing Cuts Reshape City Governments in Michigan."

44. Laina Stebbins, "Local Governments Applaud Legislature's Proposed Revenue
   Sharing Boost," Capital News Service, April 21, 2017, http://news.jrn.msu.edu/2017
   /04/local-governments-applaud-legislatures-proposed-revenue-sharing-boost/.

45. Jonathan Oosting, "How Michigan's Revenue Sharing 'Raid' Cost Communities
   Billions for Local Services," *FJ*, March 30, 2014, https://www.mlive.com/lansing
   -news/2014/03/michigan_revenue_sharing_strug.html.

46. Michigan State University Extension Center for Local Government Finance and
   Policy, *A Review of Michigan's Local Financial Emergency Law*, April 21, 2017, https://
   www.canr.msu.edu/uploads/resources/pdfs/michigan_em_law_review.pdf.

47. "Michigan EFM Outsources Water Treatment to Company Facing 26 Felony Clean
   Water Act Violations," *Eclectablog*, June 9, 2011, https://www.eclectablog.com/2011
   /06/breaking-michigan-efm-outsources-water.html.

48. Gus Burns, "Former Highland Park EFM Arthur Blackwell Won't Serve Jail Time,"
   FJ, April 17, 2013, https://www.mlive.com/news/detroit/2013/04/former_highland
   _park_efm_arthu.html.

49. Pauling, Dawn et al. v. Governor Rick Snyder et al., 2:16-cv-11263-SFC-APP, US Eastern District of Michigan, Southern Division, filed April 6, 2016, 29, 32, http://voiceofdetroit.net/wp-content/uploads/DPS-lawsuit-v-Snyder.compressed.pdf.
50. *Local Financial Stability and Choice Act*, 436 of 2012, MCL 141.1541 to 141.1575.
51. Bellant v. Snyder, 2:17-cv-13887, District Court, E. D. Michigan, 22 (2017), accessed February 3, 2019, https://ccrjustice.org/sites/default/files/attach/2017/12/Complaint%20for%20Declaratory%20Relief%202017-12-01.pdf.
52. Pauling, Dawn et al. v. Governor Rick Snyder et al., 92–93.
53. Pauling, Dawn et al. v. Governor Rick Snyder et al., 7, 94.
54. Waid v. Snyder, 5:16-cv-10444, District Court, US Eastern District of Michigan, Amended Complaint, entered September 12, 2016, 23.
55. Flint Democracy League et al., *Petition Alleging Violations of the Human Rights of Citizens of Flint*, 8–9.
56. Albert Bandura, "Self-Efficacy Mechanism in Human Agency," *American Psychologist* 37, no. 2 (1982): 122–47; Albert Bandura, "Exercise of Human Agency through Collective Efficacy," *Current Directions in Psychological Science* 9, no. 3 (2000):75–78.
57. Jacinta M. Gau, "Unpacking Collective Efficacy: The Relationship between Social Cohesion and Informal Social Control," *Criminal Justice Studies* 27, no. 2 (2014): 210–25; Robert J. Sampson and Corina Graif, "Neighborhood Social Capital as Differential Social Organization: Resident and Leadership Dimensions," *American Behavioral Scientist* 52, no. 11 (2009): 1579–1605.
58. Daniel Brisson, "Neighborhood Social Cohesion and Food Insecurity: A Longitudinal Study," *Journal of the Society for Social Work and Research* 3, no. 4 (2012): 268–79; Daniel Brisson, Susan Roll, and Jean East, "Race and Ethnicity as Moderators of Neighborhood Bonding Social Capital: Effects on Employment Outcomes for Families Living in Low-Income Neighborhoods," *Families in Society* 90, no. 4 (2009): 368–74; Catherine E. Ross and Sung J. Jang, "Neighborhood Disorder, Fear, and Mistrust: The Buffering Role of Social Ties with Neighbors," *American Journal of Community Psychology* 28 (2000): 401–20.
59. Minsoo Jung, Leesa L. Lin, and Kasisomayajula Viswanath, "Associations between Health Communication Behaviors, Neighborhood Social Capital, Vaccine Knowledge, and Parents' H1N1 Vaccination of Their Children," *Vaccine* 31, no. 42 (2013): 4860–66.
60. Michelle Anderson, "The New Minimal Cities," *Yale Law Journal* 123, no. 5 (2014): 1118–227; Pascale Joassart-Marcelli, "Ethnic Concentration and Nonprofit Organizations: The Political and Urban Geography of Immigrant Services in Boston, Massachusetts," *International Migration Review* 47, no. 3 (2013): 730–72; Pascale M. Joassart-Marcelli, Juliet A. Musso, and Jennifer R. Wolch, "Fiscal Consequences of Concentrated Poverty in a Metropolitan Region," *Annals of the Association of American Geographers* 95, no. 2 (June 2005): 336–56; J. Wolch and S. Dinh, "The New Poor Laws: Welfare Reform and the Localization of Help," *Urban Affairs Quarterly* 22 (2001): 482–89.
61. C. Gillette, "Dictatorships for Democracy: Takeovers of Financially Failed Cities," *Columbia Law Review* 114, no. 6 (2014): 1373–462; K. Bowman, "State Takeovers of

School Districts and Related Litigation: Michigan as a Case Study," *Urban Lawyer* 45, no. 1 (2013): 1–19.

62. Kevin D. Hart, Stephen J. Kunitz, Ralph R. Sell, and Dana B. Mukamel, "Metropolitan Governance, Residential Segregation, and Mortality among African Americans," *American Journal of Public Health* 88, no. 3 (1998): 434–38; J. P. Allen and E. J. Turner, "Spatial Patterns of Immigrant Assimilation," *Professional Geographer* 48 (1996): 140–55.

63. Anderson, "New Minimal Cities."

64. Jiquanda Johnson, "City Pays Councilman $4,500 after He Was Taken Out of Meeting in Handcuffs," *FJ*, November 9, 2016, https://www.mlive.com/news/flint/2016/11/meeting_removal_leads_to_damag.html.

65. Edwards's sampling technique rightfully guided his team to look for lead problems in places that experts expect them to be. This includes older homes in poor areas with high percentages of Black Americans. While this method generated important data for Flint residents, this sampling technique could not be used to determine whether the problem was concentrated in suspected areas or the extent to which the problem could be found in other spaces in the community.

66. To identify trends and population correlates of lead exposure among children under eleven years old in Flint, I calculated descriptive statistics, compared blood lead levels across the strata of covariates (e.g., census tracts, self-reported racial and ethnic identity, and median income household income at the parcel level), and explored relationships among predictors. Children residing in Flint, as identified by the MDHHS variable "city," were identified as lead poisoned when they had blood lead levels at or above 5 micrograms per deciliter ($\mu g/dL$). Since I included one blood test per child in this analysis, only the highest blood lead result was used for children with multiple blood lead tests. Age and time/period variables are transformations of variables (e.g., date of birth, date specimen collected) provided by the MDHHS. Accordingly, physical addresses in the private health data were used to calculate distance variables and to examine associations between lead exposure and proximity to neighborhood blight, which includes in this study brownfield sites, water main breaks, and high water lead levels discovered in the city. With this data, I examined demographic patterns within Flint by sorting trends by race, age, and class and calculating descriptive statistics and computing distributions. I also performed bivariate analyses of the relationships between blood level concentration and potential exposure correlates.

   The analysis of follow-up testing in chapter 5 is an effort to specify the extent to which children with lead poisoning in Flint received a second blood lead testing within three months, as required by federal and state law. I restricted the analysis of follow-up testing trends to preschool children (target population of retesting efforts) with lead poisoning to 2013–15 in order to ensure that health providers were utilizing the same definition of blood lead poisoning. I used logistic regression to examine the associations between follow-up testing and demographic indicators (race, income at parcel level, blood lead level, Medicaid status, moved after first blood lead test, and age of housing) while controlling for pertinent child and family characteristics (sample type, sex, age, and ethnicity).

67. Jacob Abernethy, Cyrus Anderson, Chengyu Dai, Arya Farahi, Linh Nguyen, Adam Rauh, Eric Schwartz, et al., "Flint Water Crisis: Data-Driven Risk Assessment via Residential Water Testing," presented at the Data for Good Exchange 2016, arXiv:1610.00580v1, accessed September 30, 2016.

68. I explored the distribution of repairs made by phase (1–3 v. 4) and zip codes and identified correlates of full (both) as well as partial (public and private) pipe replacements. I also performed bivariate analyses to assess relationships between replacement type and potential correlates. I utilized binary logistic regression to calculate adjusted odds ratios to examine the associations between replacement types and Flint area zip codes, controlling for demographic indicators including mean household income, percentage of homes built before 1950, and percentage of vacant homes. Logistic regression models were also utilized to determine the likelihood of full service line replacements, controlling for risk factors associated with elevated water lead levels in Flint, which include housing stock age (home built before 1950) and a poverty indicator (percentage below poverty level). This analysis also controls for the percentage of owner-occupied homes, property owner type, and zip code.

CHAPTER 1

1. Mathew Forstater, "From Civil Rights to Economic Security: Bayard Rustin and the African-American Struggle for Full Employment, 1945–1978," *International Journal of Political Economy* 36, no. 3 (2007): 63–74; Joe Feagin, "Slavery Unwilling to Die: The Background of Black Oppression in the 1980s," *Journal of Black Studies* 17, no. 2 (1986): 173–200.

2. Peggy Seigel, "Pushing the Color Line: Race and Employment in Fort Wayne, Indiana, 1933–1963," *Indiana Magazine of History* 104, no. 3 (2008): 241–76.

3. Robert R. Gioielli, "The Breakdown of the City," in *Environmental Activism and the Urban Crisis: Baltimore, St. Louis, Chicago* (Philadelphia: Temple University Press, 2014), 11–37.

4. Quintin Johnstone, "The Federal Urban Renewal Program," *University of Chicago Law Review* 25, no. 2 (1958): 301–54.

5. Deborah Wallace and Rodrick Wallace, *A Plague on Your Houses: How New York Was Burned Down and National Public Health Crumbled* (London: Verso, 2001).

6. Joe T. Darden, Richard Child Hill, June Thomas, and Richard Thomas, *Detroit: Race and Uneven Development* (Philadelphia: Temple University Press, 1987); US National Advisory Commission on Civil Disorders, *The Kerner Report: The 1968 Report of the National Advisory Commission on Civil Disorders* (New York: Pantheon, 1988).

7. Charles Monroe Haar, *Between the Idea and the Reality: A Study in the Origin, Fate, and Legacy of the Model Cities Program* (Boston: Little, Brown, 1975); Bernard J. Frieden and Marshall Kaplan, *The Politics of Neglect: Urban Aid from Model Cities to Revenue Sharing* (Cambridge, MA: MIT Press, 1975).

8. Lawrence D. Brown and Bernard J. Frieden, "Rulemaking by Improvisation: Guidelines and Goals in the Model Cities Program," *Policy Sciences* 7, no. 4 (1976): 455–88; Charles E. Olken, "Economic Development in the Model Cities Program," *Law and Contemporary Problems* 36, no. 2 (1971): 205–26; Judson L. James, "Federalism and the Model Cities Experiment," *Publius* 2, no. 1 (1972): 69–94.

9. James, "Federalism and the Model Cities Experiment."

10. Erasmus Kloman, "Citizen Participation in the Philadelphia Model Cities Program: Retrospect and Prospect," *Public Administration Review* 32 (1972): 402–8; John H. Strange, "Citizen Participation in Community Action and Model Cities Programs," *Public Administration Review* 32 (1972): 655–69.

11. A. C. Findlay, "Population of Flint," Compiled Studies of Flint Institute of Research and Planning (1938), 3–5, Alexander C. Findlay papers, Flint Institute of Research and Planning—Compiled Studies, 1936–1939 folder, GHCC.

12. Pierce F. Lewis, "Geography in the Politics of Flint" (PhD diss., University of Michigan, 1959).

13. Lewis, 572.

14. Frank J. Corbett and Arthur J. Edmunds, *Some Characteristics of Real Property Maintenance by Negro Home Occupants* (Flint: Urban League of Flint, 1954).

15. Andrew R. Highsmith, *Demolition Means Progress: Flint, Michigan, and the Fate of the American Metropolis* (Chicago: University of Chicago Press, 2015), 421–22.

16. James L. Rose (Director of Housing, Michigan Civil Rights Commission), "Bulldozers vs. People," November 1966, OBP, box 7.

17. M. McLaughlin, *The Long, Hot Summer of 1967: Urban Rebellion in America* (New York: Palgrave Macmillan, 2014).

18. Dominic Adams, "See How Riots in Detroit 50 Years Ago Spread to Flint," *FJ*, July 24, 2017, updated May 20, 2019, https://www.mlive.com/news/flint/2017/07/see_how_riots_in_detroit_50_ye.html.

19. Raymond V. Meagher, "Disorder in Flint Subsides Except for the Fire Problem," *FJ*, July 27, 1967, https://www.blackpast.org/african-american-history/flint-michigan-riot-1967/.

20. Highsmith, *Demolition Means Progress*, 318–19, 326.

21. *Model Cities News*, October 1972, OBP, box 10.

22. Model Cities Program, "Resident Organization: Program Administration Budget, Genesee County CDA," November 26, 1969, OBP, box 11.

23. "Genesee County Model Cities Program," 1968; Henry Horton, "CDA Planning Progress Report," August 10, 1968; Frances Leonard, "Memo RE: Genesee County Model Cities Application," June 11, 1969; and "Model Cities Planning Progress Report," January 10, 1969, all in OBP, box 11.

24. Letter from John W. Mack, Executive Director, the Urban League of Flint, May 11, 1967, OBP, box 11.

25. Letter from David Leonard to DeHart Hubbard, Intergroup Relations Officer, HUD, May 9, 1967, OBP, box 11.

26. Leonard to Hubbard; "Model Cities Planning Progress Report, Period: September 1, 1968, to January 10, 1969," OBP, box 11.

27. US Department of Housing and Urban Development, "HUD Model Cities Contract Offered Genesee County, Mich.," October 15, 1969, OBP, box 11.

28. "What Is Model Cities," OBP, box 11.

29. *Model Cities News*, May 1972, OBP, box 10.

30. *Model Cities News*, August 1970, OBP, box 10; July 1970, OBP, box 10.

31. *Model Cities News*, May–June 1971, OBP, box 10

32. *Model Cities News*, December 1970, OBP, box 10.

33. *Model Cities News*, May–June 1971.

34. *Model Cities News*, October 1972, OBP, box 10.

35. "Biography: Olive Beasley Papers," University of Michigan–Flint, GHCC, accessed December 19, 2019, https://www.umflint.edu/archives/olive-beasley-papers.

36. Olive Beasley [to James Rose], "Model Cities Quarterly Report," August 3, 1970, OBP, box 8.

37. Beasley, "Model Cities Quarterly Report."

38. Urban League of Flint Housing Center, "Report on the Urban League of Flint Mini-Conference on Housing," December 17, 1974, Cox Papers, box 4, GHCC.

39. P. W. Dawkins (City of Flint, Department of Community Development Technician), "Fair Housing in Flint Memorandum," August 17, 1978, Cox Papers, box 2, GHCC.

40. "Project Budget" (City of Flint), 115, 117, OBP, box 11.

41. "Community Development Block Grant Allocations" (City of Flint, Department of Community Development), tables A1–A5, OBP, box 11.

42. "Flint Schools to Observe Black History and Brotherhood Weeks" (News Communications, Flint Board of Education), February 10, 1974, Holt Papers, box 5, GHCC.

43. Olive Beasley, "Flint Schools—Desegregation," June 20, 1973; and Flint Community Schools, "Project Onward and Upward: Middle Cities Program," February 1971, Holt Papers, box 5, GHCC.

44. Letter to Dr. Peter L. Clancy (Superintendent, Flint Community Schools) from US Department of Health, Education, and Welfare, August 29, 1975, Holt Papers, box 5, GHCC.

45. NAACP Flint Board of Education, "Fact Sheet: Holman et al. vs. Flint Community Schools et al.," March 23, 1976, Holt Papers, box 5, GHCC; "Letter to Olive R. Beasley from Edgar Holt Regarding Flint Public School Issues," April 22, 1969, NAACP Flint Board of Education, Holt Papers, box 5, GNCC; NAACP Flint Branch—Beecher School District, "Note to Beecher Citizens from the Flint Branch of NAACP Re School Desegregation," September 3, 1973, Holt Papers, box 8, GHCC.

46. NAACP Flint Board of Education, "Letter of Determination by Department of Health, Education, and Welfare," August 29, 1975, Holt Papers, box 5, GHCC.

47. Community education emerged in Flint in the 1930s, partially in response to labor concerns brought on by the birth of the United Auto Workers and by significant population shifts that threatened the cohesiveness and productivity of this relatively homogeneous community. To counter the union's democratizing and galvanizing influence—in which workers were encouraged to think of themselves as a united working class—General Motors executive and arch segregationist Charles

Stewart Mott joined forces with educator Frank Manley. Their resulting "community education" program transformed Flint public school facilities into vibrant civic spaces. Educational opportunities were expanded to encourage the development of a more engaged and socially connected community. When African Americans settled in previously segregated neighborhoods, racial tensions emerged. In response to these tensions, community education leaders campaigned to maintain racially segregated schools and created policies that institutionalized and encouraged racial separation and performance disparities across area schools.

48. NAACP Flint Board of Education, "Letter of Determination by Department of Health, Education, and Welfare," 248.

49. NAACP Flint Board of Education, "Letter of Determination by Department of Health, Education, and Welfare," 251.

50. NAACP Flint Board of Education, "Letter of Determination by Department of Health, Education, and Welfare," 258.

51. Michigan Civil Rights Commission, "Proposals for Dealing with Allegations of Police Misconduct in Flint, Michigan," August 31, 1967, OBP, box 10.

52. Preliminary Complaint Report: Walter Mitchell, NAACP, Flint, Holt Papers, box 9, GHCC.

53. Preliminary Complaint Report: Walter Mitchell.

54. Thomas E. Sagendorf (Director, Interfaith Action Council of Greater Flint), "An Account" (claims against Flint Police Department, July 1, 1969), OBP, box 18.

55. Field Representative Notes: William Brown, NAACP, Flint, OBP, box 14.

56. Field Representative Notes: William Brown.

57. "Response to Ad Hoc Committee Requests" (City of Flint, Office of the Mayor, Honorable James W. Rutherford), January 14, 1976, OBP, box 23.

58. Olive R. Beasley (District Executive), "Request for Information—Flint Police Cases," July 16, 1971, OBP, box 14.

59. Olive R. Beasley (District Executive), "Flint Police–Community Tensions," January 7, 1972, OBP, box 14.

60. Beasley, "Flint Police–Community Tensions"; "Field Representative's Notes" (claims against Flint Police Department, November 10, 1971, January 2, 1970, and December 29, 1969), OBP, box 18.

61. "Request for Information—Flint Police Cases" (claims against Flint Police Department, July 16, 1971), OBP, box 18.

62. "Michigan Civil Rights Commission News Release" (claims against Flint Police Department, September 25, 1970), OBP, box 18.

63. "Flint Police–Community Tensions" (claims against Flint Police Department, January 7, 1972), OBP, box 18.

64. James A. Williams, "To Transform the Inner City: Tucson's Model Cities Program, 1969–1975," *Journal of Arizona History* 52, no. 2 (2011): 143–68.

65. Daniel Elazar, "From the Editor of *Publius*: Courts, Communities, and the New Federalism," *Publius* 8, no. 4 (1978): 13; David Walker, "The Advent of an Ambiguous Federalism and the Emergence of New Federalism VIII," *Public Administration Review* 56, no. 3 (1996): 271–80.

66. Planning and Programing Division, Department of Community Development, Flint, Michigan, "City of Flint Housing Assistance Plan" (CDBG Years 8, 9, 10 — Fiscal Years 1983, 1984, and 1985) and "Community Development Block Grant: The Historical Perspective," 88–94, OBP, box 11.

67. "Condition of Housing in Flint: Focus—The Black Community," Urban League Mini-Conference on Housing, November 1974, 5, Cox Papers, box 4, GHCC.

68. Clemens Zimmermann, *Industrial Cities: History and Future*, Interdisciplinary Urban Research (Frankfurt, Germ.: Campus Verlag, 2013), 157.

69. Letter from Olive Beasley et al. to Richard Wilberg (Flint Department of Community Development, Affirmative Action), June 7, 1973, OBP, box 21, 3.

70. City Ordinance #2008 (passed 5–4 in October 1967). It was designed to outlaw housing discrimination and met with opposition from segregationists, who secured the needed signatures to put the measure on the ballot, but the special election upheld the city's fair housing policy.

71. Fair Housing in Flint memorandum (letter from F. W. Dawkins, City of Flint, Michigan, to Jack Litzenberg), August 17, 1978, Flint Archives (Flint DCD Fair Housing Program), Cox Papers, box 2, GHCC.

72. Highsmith, *Demolition Means Progress*, 347.

73. Highsmith, 332.

74. "Condition of Housing in Flint: Focus Black Community," Urban League of Flint Housing Center, p. 1, ECHO [Evidence for Community Health Organization] Program collection, box 3, GHCC.

75. Urban League Housing Center, "Condition of Housing in Flint: Focus—The Black Community" (Urban League Mini-Conference on Housing), November 1974, Flint Archives, Cox Papers, box 4, GHCC; Lonnie Fouty, "Proposal for Flint, Michigan," September 29, 1974 (NAACP housing information), Flint Archives, Holt Papers, box 5, GHCC.

76. "Model Cities–Quarterly Report Memorandum," letter to James Rose, Housing Program from Olive Beasley, p. 1, August 3, 1970, Beasley box 7. C.R.C. Housing Department Memoranda, 1970–1974, GHCC.

77. S. Weissman, "The Limits of Citizen Participation: Lessons from San Francisco's Model Cities Program," *Western Political Quarterly* 31, no. 1 (1978): 32–47; Leanne C. Serbulo and Karen J. Gibson, "Black and Blue: Police–Community Relations in Portland's Albina District, 1964–1985," *Oregon Historical Quarterly* 114, no. 1 (2013): 6–37.

CHAPTER 2

1. Karen Gatlin, "Report on the Opening of AutoWorld," WJRT ABC12 News, July 4, 1984; "AutoWorld: The Failed Theme Park That Was Supposed to Save Flint, Michigan," *Ostrich Report*, February 25, 2019.

2. Ron Fonger, "30 Years, 30 Pictures: See the History of Autoworld Three Decades after It Opened," *FJ*, July 3, 2014, https://www.mlive.com/entertainment/flint/2014/07/autoworld_in_restrospect_30_ye.html; Tom Hundley, "Michigan Theme Park Still Idling," *Chicago Tribune*, May 8, 1988, https://www.chicagotribune.com/news/ct-xpm-1988-05-08-8803150188-story.html.

3. George F. Lord and Albert C. Price, "Growth Ideology in a Period of Decline: Deindustrialization and Restructuring, Flint Style," *Social Problems* 39, no. 2 (1992): 155–69; Ron Fonger, "'They Needed More Thrill': Looking Back at 25th Anniversary of Flint Legend, AutoWorld," *FJ*, July 3, 2009 https://www.mlive.com/news/flint /2009/07/they_needed_more_thrill_lookin.html.

4. United States v. Malcolm Wilson, US Court of Appeals, Sixth Circuit, 972 F.2d 349, No. 91–1510, July 28, 1992.

5. US Senate Committee on Governmental Affairs, Permanent Subcommittee on Investigations, *Drugs and Violence: The Criminal Justice System in Crisis* (Washington, DC: Government Printing Office, 1989), 393.

6. Jennifer Carlson, *Citizen-Protectors: The Everyday Politics of Guns in an Age of Decline* (New York: Oxford University Press, 2015).

7. "Meeting the Problem of Infant Mortality," February 19, 1983, 12, James Blanchard Papers, box 104, BHL.

8. Robert Longstaff, "Environment Unit 'Race' Likely," *FJ*, November 16, 1972.

9. "New Priorities for a New Decade in Environmental Protection," 5, John Engler Papers, box 11, BHL.

10. Greg Braknis, "Flint GM Plan among 19 Facing Pollution Suits," *FJ*, November 18, 1986; Greg Braknis, "Flint-Area GM Plant among 17 Settling State Pollution Suits," *FJ*, November 19, 1986; Greg Braknis, "Plant Has History of Pollution Woes," *FJ*, November 18, 1986.

11. During this time, General Motors also contributed to air pollution in the Flint area. This was especially true near the industrial complex off Stewart Avenue and Selby Street, where the company removed oils from the industrial wash water. Although the smell was strongest on the north side of Flint, this odor was also known to travel east and was described as "sulfuric and skunky" (Teri Banas, "Some Residents Raising Stink over GM Plant Odor," *FJ*, February 21, 1990). The odor itself was considered harmless by the Department of Natural Resources, defined as posing "no injurious effects to human health or safety" at the time. Many residents, however, complained that they could not cook or spend time outside—or even have their windows open—during the summer because the stench from the treatment plant was so strong. Following various environmental assessments and cleanups at GM plants like Buick City and Flint West (including its flagship Chevy in the Hole factory on the Flint River) during the early to mid-2000s, GM for all intents and purposes left the city hanging. The company sold its flagship factory to Flint for one dollar and left behind a legacy, as told by the soil and groundwater of the abandoned Flint plants, filled with heavy metals (including lead, mercury, arsenic, volatile organic compounds, and other toxic substances).

12. "BEN Evaluation Study" (memorandum by Gregory McKenzie, P. D. Administrator, addressed to the City of Flint), August 20, 1981, 5, Cox Papers, box 3, GHCC.

13. Ladislas Segoe and Associates, "Comprehensive Master Plan of Flint, Michigan, and Environs," April 30, 1959, [Cincinnati], 1960, 13, University of Michigan Library.

14. Ladislas Segoe and Associates, "Comprehensive Master Plan of Flint, Michigan, and Environs," table 3.

15. "BEN Evaluation Study," 10–11.
16. "BEN Evaluation Study," 11–12.
17. "BEN Evaluation Study," 4.
18. "BEN Evaluation Study."
19. "BEN Evaluation Study," 49.
20. Michael J. Riha, "Air in Area Fails to Meet Standards," *FJ*, February 3, 1978; Ed Conaway, "Speakers Oppose Air Pollution Plan," *FJ*, February 18, 1979.
21. Ed Conaway, "City of Flint Is Unable to Control Air Pollution," *FJ*, January 20, 1978.
22. "Proposed Rules," *Federal Register* 63, no. 13 (January 21, 1998), https://www.govinfo.gov/content/pkg/FR-1998-01-21/pdf/98-1249.pdf.
23. Stuart Bauer, "Berlin and Farro: First Owner of House Near Toxic Site Got Cancer," *FJ*, January 29, 2008, https://www.mlive.com/flintjournal/newsnow/2008/01/berlin_and_farro_first_owner_o.html .
24. Julie Morrison and Ryan Garza, "Taking Another Look at Berlin & Farro," *FJ*, January 26, 2008, https://www.mlive.com/flintjournal/newsnow/2008/01/taking_another_look_at_berlin.html.
25. Ronald Lynn Voelker obituary, Swartz Funeral Home, accessed December 20, 2019, https://www.swartzfuneralhomeinc.com/obituary/5963459.
26. Shyra Lynn Voelker obituary, Legacy.com, accessed December 20, 2019, https://www.legacy.com/obituaries/name/shyra-voelker-obituary?pid=178274572.
27. Jeff Rauschert, "Berlin and Farro: First Owner of House Near Toxic Site Got Cancer," *FJ*, January 29, 2008, https://www.mlive.com/flintjournal/newsnow/2008/01/berlin_and_farro_first_owner_o.html.
28. "Environmental Workshop Told of Block Clubs' Value," *FJ*, October 1, 1971.
29. Mike Stobbe, "Landfills Continue to Lead Worry List about Environment," *FJ*, December 30, 1989.
30. Roger Van Noord, "Inspector Says Flint Breaks Own Pollution Laws," *FJ*, July 2, 1972.
31. Rhonda Sanders, "Residents Find Little Amusing at Old Park," *FJ*, September 1, 1981; Stobbe, "Landfills Continue to Lead Worry List about Environment."
32. Lee Bergquist, "Flint Council Gripes, but Garbage Pickup Still Will Be Biweekly," *FJ*, September 19, 1981.
33. Lee Bergquist, "Flint Council Gripes, but Garbage Pickup Still Will Be Biweekly," *FJ*, September 19, 1981.
34. Chris Christoff, "Weekly Trash Pickups to Start," *FJ*, December 9, 1982.
35. By 1993, garbage pickup in Flint was privatized after Mayor Woodrow Stanley and the city council concluded that going private would alleviate the increasing number of calls received from Flint residents as well as the equipment shortages and the lack of personnel that were hindering the city's capacity to response to medical emergencies (Todd Seibt, "Officials Say Paramedic Tax Needed Because of More Calls," *FJ*, May 29, 1986). Efforts to restructure medical units, an initiative that involved the Flint Fire Department, did not improve the efficiency of paramedic teams (Dan Shriner, "Emergency Medical Units Restructured, Streamlined," *FJ*, July 7, 1992).

Although residents feared potential increases in the cost of paramedic services and the loss of services to poor people without health insurance, the city contracted with Pulse to transport patients between hospitals and on ambulance runs, allowing it to be the first privately owned paramedic service in the City of Flint. See Rhonda S. Sanders, "Clean Up Crew Finds Its Arms Too Short to Box with City Hall," *FJ*, August 29, 1981; and Laura A. Bischoff, "Private Pickup: $3 Million Savings," *FJ*, June 11, 1993.

36. Greg Braknis, "Genesee County Reports Gusher of Leaking Underground Tanks," *FJ*, August 28, 1988.

37. Robert Lewis, "Study Indicates High Lead Level in Flint's Soil," *FJ*, May 27, 1972.

38. Lewis, "Study Indicates High Lead Level in Flint's Soil."

39. Braknis, "Genesee County Reports Gusher of Leaking Underground Tanks"; Wendy Wendland, "Station Operator Faces Loss of Life's Savings over Gas Spill Cleanup," *FJ*, January 19, 1992; Mike Stobbe, "State Sues Action Auto Alleging Leaking Tanks," *FJ*, May 24, 1990.

40. Owners were correct to assume that they had access to state money to fix the problems caused by leaking underground storage tanks. Beginning February 15, 1990, the State of Michigan announced that it would make grants from the $300 million leaking tank cleanup fund available to gas station owners to deal with spills that occurred after July 18. The state was aware that a large percentage of tank owners (almost 75 percent) would need funding, since many were in violation of environmental laws and lacked the financial resources to monitor and clean up leaks from old tanks. Although the idea of the cleanup fund was progressive and made Michigan a leader in the war on leaking underground storage tanks, in reality many gas station owners never gained access to these funds. As a result of the rising cost of maintaining underground tanks and the lack of resources available to assist tank owners, many of Michigan's gas stations went out of business. Jeff Alexander, "Cleanup Fund Barred at Any Leaky Tank Sites," *FJ*, January 8, 1990.

41. Tim Doran, "Well Woes Mean Tapping New Routine," *FJ*, July 28, 1989; Greg Braknis, "Gasoline Leak? Family's Tainted Water Raises Health Concerns," *FJ*, March 16, 1989; Greg Braknis, "N. Term Resident's Hopes Dashed as Well Water Found Contaminated," *FJ*, April 27, 1989.

42. Mike Stobbe, "Gas Leaks from Tanks Buried Underground Alarm Health Officials," *FJ*, June 6, 1990; Mike Stobbe, "Fears Linger," *FJ*, April 1, 1990.

43. House Speaker v. Governor, 443 Mich. 560, 572 (506 N.W.2d 190 1993).

44. United Press International, January 30, 1992, and January 6, 1993; *DN*, September 29, 1993.

45. "Environmental Employees Say They Can't Do Their Jobs: DEQ Survey Flunks Engler Policies," October 21, 1998, Michigan Environmental Council Records, box 21, BHL; Bob Gwizdz, "DNR's Best and Brightest Bailing Out," *Ann Arbor News*, January 23, 1993.

46. "Environmental Employees Say They Can't Do Their Jobs."

47. Environmental Working Group, *Prime Suspects: The Law Breaking Polluters America Fails to Inspect* (Washington, DC: Author, 2000).

48. "Scorched Earth: An Update on the Rapid Retreat from Environmental Protection under Governor John Engler," June 3, 1999, Michigan Environmental Council Records, box 21, BHL; "Generations to Come Will Reap the Bitter Fruit of DEQ's Dereliction of Duty," October 5, 2000, Michigan Environmental Council Records, box 21, BHL.

49. "Dereliction of Duty: How the Department of Environmental Quality Endangers Michigan's Environment and Public Health," October 5, 2000, p. 27, Michigan Environmental Council Records, box 21, BHL.

50. "USEPA/Select Steel Press Conference," September 2, 1998, John Engler Papers, box 128, BHL.

51. Kary L. Moss, "Environmental Justice at the Crossroads," *William and Mary Environmental Law and Policy Review* 24, no. 1 (2001): 35–66.

52. David Mastio, "EPA Inquiry May Kill Flint Steel Mill," *DN*, August 26, 1998; David Mastio, "Mostly Whites Live near Proposed Mill Site," *DN*, August 27, 1998; David Mastio, "Flint Divided, Dismayed as EPA Crusade Falters," *DN*, December 28, 1998.

53. "Brief in Support of Plaintiffs' Motion for Injunctive Relief," July 15, 1995, 12, SFPC, box 1, folder 9.

54. "Petition of American Lung Association of Michigan to Environmental Appeals Board regarding the Genesee Power Station," January 5, 1993, SFPC, box 1, folder 1.

55. "Brief in Support of Plaintiffs Motion for Injunctive Relief," 19–20.

56. "Petition of American Lung Association of Michigan to Environmental Appeals Board regarding the Genesee Power Station," 3.

57. "Response to Petitions for Review of PSD Permit of Genesee Power Station Limited Partnership," February 24, 1993, 45–46, SFPC, box 1, folder 2.

58. "Deposition of Dr. Stuart Bannerman," February 6, 1997, SFPC, box 1, folder 14.

59. Letter to Valdas Admakus, Administrator, EPA Region 5, from the Guild Law Center, July 6, 1994, SFPC, box 1, folder 2.

60. "Brief in Support of Plaintiffs Motion for Injunctive Relief," 22–23.

61. Letter to Carol Browner, EPA Administrator, Re: Failure of US EPA to Comply with 40 CFR § 7 in Processing, Investigating and Resolving Administrative Complaints under Title VI, October 14, 1996, 9, SFPC, box 2, folder 83.

62. Letter to Carol Browner.

63. "Joint Petition to EPA to Re-open Select Steel Investigation," 12, SFPC, box 2, folder 82.

64. Letter from Lilian Dorka, EPA, to Heidi Grether, MDEQ, January 19, 2017, https://www.epa.gov/sites/production/files/2017–01/documents/final-genesee-complaint-letter-to-director-grether-1-19-2017.pdf.

65. "Brief in Support of Plaintiffs Motion for Injunctive Relief," 40.

66. "Transcript of Hearing, NAACP-Flint Chapter v. Engler et al.," April 29, 1997, 82, and "Deposition of Russell J. Harding," October 17, 1996, SFPC, box 1, folder 16.

67. J. W. Bulkley, R. Y. Demers, D. T. Long, G. T. Wolff, and K. G. Harrison, *The Impacts of Lead in Michigan* (Lansing: Michigan Environmental Science Board, 1995); David Mastio, "Governor Will Use Flint Press Conference to Denounce

Environmental Justice Rules," *DN*, September 2, 1998, B1; Tom Wickham, "'Environmental Justice' Dispute Harmful—Engler" (editorial), *FJ*, November 11, 1998, A5; "Our Views: Round 2 over Steel Mill Could Be Disastrous," *FJ*, November 3, 1998, A6; David Mastio, "EPA Aids Activist Groups: Agency Gives $10 Million to Organizations Pushing Its Environmental Justice Campaign," *DN*, June 5, 1998, A6.

68. "Select Steel Appeal Number PSD 98–21, Response to Petition of the SFPC," August 19, 1998, 1, SFPC, box 2, folder 79.

69. Dan O'Brien memo to Hien Nguyen, "Review of Permit Conditions Regarding Mercury Emissions, Permit #579–97," March 31, 1998, SFPC, box 2, folder 94.

70. W. Charles McIntosh to Ann Goode, Director of OCR, September 30, 1998, SFPC, box 2, folder 90.

71. Anderson Economic Group, "Demographics of 4-Tract Area Surrounding Plant, Genesee County," August 26, 1998, John Engler Papers, box 128, BHL.

72. "Select Steel Appeal Number PSD 98–21, Response to Petition of the SFPC," 2.

73. "Select Steel Appeal Number PSD 98–21, Response to Petition of the SFPC," 6; "Select Steel Press Conference," September 1, 1998, John Engler Papers, box 128, BHL.

74. "Select Steel Appeal Number PSD 98–21, Response to Petition of the SFPC," 6.

75. Alex Sagady, "E-M:/Engler News Release on Select Steel Case," Alex J. Sagady and Associates, September 3, 1998, http://www.great-lakes.net/lists/enviro-mich/1998-09/msg00016.html; David Mastio, "Governor Will Use Flint Press Conference to Denounce Environmental Justice Rules," *DN*, September 2, 1998, B1; Paul Connolly, "Environmental Justice: Michigan Governor Blames EPA Policy for Company Decision to Scrap Factory Plan," 29 Env't Rp. (BNA), at 995–96 (September 18, 1998); Wickham, "'Environmental Justice' Dispute Harmful"; "Our Views: Round 2 over Steel Mill Could Be Disastrous"; Mastio, "EPA Aids Activist Groups."

76. "Environmental Appeals Board Order Denying Review," September 10, 1998, SFPC, box 2, folder 80.

77. Bradford C. Mank, "Title VI," in *The Law of Environmental Justice: Theories and Procedures to Address Disproportionate Risks*, ed. Michael B. Gerrard and Sheila Foster, 23–66 (Chicago: American Bar Association, 2008); Moss, "Environmental Justice at the Crossroads."

78. "Area Loses Steel Mill, Report Says," *DN*, March 2, 1999; David Mastio, "EPA Race Policy Costs Flint Plant," *DN*, March 2, 1999.

79. Bill Johnson, "Save Detroit's Development and Minorities from Eco-Justice," *DN*, June 12, 1999.

80. "MDEQ Response to the SFPC Title VI Complaint of June 9, 1998 regarding Select Steel," September 16, 1998, 3–4, SFPC, box 2, folder 81.

81. "MDEQ Response to the SFPC Title VI Complaint of June 9, 1998," 5.

82. "MDEQ Response to the SFPC Title VI Complaint of June 9, 1998."

83. Transcript of testimony of Lillian Robinson, United for Action, Justice, and Environmental Safety, before the National Environmental Justice Advisory Council, December 9, 1998, SFPC, box 2, folder 85.

84. Robinson transcript.

85. Robinson transcript.

86. David Poulson. 1993. "Engler plan to let polluted water lie roils controversy." *FJ*. June 6, 1993. FJ Archives.

87. Then-governor John Engler previously separated the state agencies in 1995.

88. "State of Michigan, Environmental Justice Plan," Michigan.gov, December 17, 2010, iv, accessed February 18, 2019, https://www.michigan.gov/documents/deq/met _ej_plan121710_340670_7.pdf.

89. "State of Michigan, Environmental Justice Plan," 23–24.

90. State of Michigan, "Executive Order 2011–1," January 4, 2011, accessed October 18, 2020, https://www.michigan.gov/documents/snyder/EO-01-2011_342039_7.pdf.

91. FCC, transcript of public comment period, April 30, 2014.

92. Quotes from FCC, transcript of public comment period, January 13, 2014.

CHAPTER 3

1. City of Flint, "City of Flint 2014 Annual Water Quality Report," https://www .cityofflint.com/wp-content/uploads/CCR-2014.pdf (August 2015).

2. Dominic Adams, "Closing the Valve on History: Flint Cuts Water Flow from Detroit after Nearly 50 Years," *FJ*, April 25, 2014, https://www.mlive.com/news /flint/2014/04/closing_the_valve_on_history_f.html.

3. Dominic Adams, "Closing the Valve on History: Flint Cuts Water Flow from Detroit after Nearly 50 Years," *FJ*, April 25, 2014, https://www.mlive.com/news /flint/2014/04/closing_the_valve_on_history_f.html.

4. Unidentified Flint Resident reported to EPA through the Safe Drinking Water Hotline, USEPA FOIA, email, October 2, 2014.

5. "Cities Sewage Facilities Are Inadequate," *FJ*, June 26, 1959.

6. "Pollution of River by Buick Is Charged," *FJ*, March 30, 1967; Lawrence R. Gustin, "River Pollution by Industry Continues, Says Water Chief," *FJ*, December 1, 1965.

7. Kenneth L. Peterson, "Local Streams Run Clean until They Meet People," *FJ*, June 14, 1970.

8. "Flint Order to Act on Creek Problem," *FJ*, May 20, 1955.

9. Marlon S. Vestal, "Statement Issued in Lansing: State Water Official Invites Talks on River Condition," *FJ*, February 18, 1960; Randolph H. Pallotta, "WRD Grants City's Appeal," *FJ*, June 1, 1960.

10. Pallotta, "WRD Grants City's Appeal"; "State Summons Flint to Pollution Hearing," *FJ*, April 29, 1960; "State Tells Flint to Halt Pollution by Early in 1963," *FJ*, July 28, 1960.

11. "Answer Filed by City in River-Pollution Suit," *FJ*, May 3, 1961.

12. "Say Chemical Fatal to Fish in Flint River," *FJ*, July 28, 1962; Marion S. Vestal, "Statement Issued in Lansing," *FJ*, February 19, 1960; Duane E. Poole, "300 Cases Listed So Far in 1961," *FJ*, August 17, 1961; Homer E. Dowdy, "Hepatitis: One Disease Traced to Pollution," *FJ*, May 28, 1961; "Flint Order to Act on Creek Problem," *FJ*, May 20, 1955.

13. "Commissioner Says Further River Pollution Steps Needed," *FJ*, June 8, 1965; "Flint River Pollution Study Asked," *FJ*, August 20, 1966; A. Allan Schmid, *Michigan Water Use and Development Problems*, Michigan State University Agricultural Experiment Station, Circular Bulletin, January 1961.

14. Kenneth Peterson, "Maintaining Quality of the Flint River Long Has Been a Problem in the Area," *FJ*, July 28, 1982; Kenneth Peterson, "Go with the Flow," *FJ*, July 29, 1982; Mike Stobbe, "River Cleaner, but Pollution Problems Persist," *FJ*, September 29, 1991; Peterson, "Local Streams Run Clean until They Meet People"; "City Is Expected to Release Survey of Pollution of Creeks Next Week," *FJ*, January 22, 1970; Kenneth Peterson, "Flint River Pollution Is Traced," *FJ*, December 24, 1967; "Check Confirms Drain-Pollution," *FJ*, August 10, 1961; Rudolph H. Pallotta, "Tests Show Bacteria Count above Normal," *FJ*, August 15, 1961.

15. "Thrall Enters Pleas of Not Guilty in Spills," *FJ*, January 16, 1979; "Thrall Fined $1000 for Polluting Sewer System Near Industrial Park," *FJ*, January 27, 1979.

16. "1900 Gallons of Gas Leaks into Flint River," *FJ*, January 30, 1979; "Gas Leak Still Poses Problems," *FJ*, January 31, 1979. Occasionally, the City of Flint was sued over its contribution to the pollution problem. In February 1978, the Flint Township voted to sue the City of Flint in order to end the city's years of pollution and force the municipality to clean up the mess. The city did not deny polluting the river. "Yes, we have been polluting the river," said A. W. DeBlaise, director of the city's public works department, to a *Flint Journal* reporter. The dispute was over the city's lag in addressing the pollution problem. A $2 million lawsuit was eventually filed against the City of Flint in March 1978 in order to compel Flint to stop polluting the river. See David Fenech, "Flint Twp. to Sue Flint over Pollution," *FJ*, February 21, 1978; and "Flint Twp. Files $2 Million Suit against Flint," *FJ*, March 22, 1978.

17. Kenneth Peterson, "Maintaining Quality of the Flint River Long Has Been a Problem in the Area," *FJ*, July 28, 1982; Kenneth Peterson, "Go with the Flow," *FJ*, July 29, 1982; Stobbe, "River Cleaner, but Pollution Problems Persist."

18. Roger P. Welshans, "Lake Water Again Flint's 'Cup of Tea,'" *FJ*, October 23, 1977.

19. Randall MacIntosh, "Flint's Water 'Safe' Despite Contamination," *FJ*, May 30, 1986.

20. Randall MacIntosh, "Flint's Water 'Safe' despite Contamination," *FJ*, May 30, 1986; David Holtz, "More Bacteria Found in Flint Water System," *FJ*, June 1, 1986; Randall MacIntosh, "How Bacteria Got in Water Still Mystery," *FJ*, June 4, 1986; Greg Braknis, "Tests Again Find Bacteria in Flint Water," *FJ*, June 12, 1986.

21. Randal MacIntosh, "Water Bacteria Problems Prompt Bottled Sales," *FJ*, June 13, 1986.

22. MacIntosh, "Water Bacteria Problems Prompt Bottled Sales."

23. Greg Braknis, "Bacteria Were Detected in Flint's Water 7 Times in 2 Years," *FJ*, June 22, 1986.

24. Greg Braknis, "Bacteria Surface Again in Flint Water," *FJ*, August 29, 1986.

25. See the following in the *Flint Journal* by Greg Braknis: "Tests Again Find Bacteria in Flint Water"; "Bacteria Were Detected in Flint's Water 7 Times in 2 Years," June 22, 1986; "Chlorine Level in Flint Water May Drop Again," August 5, 1986;

"Bacteria Surface Again in Flint Water," August 29, 1986; "Flint-Like Water 'Bugs' Have Kept Other Cities Guessing about Cure," June 22, 1986.

26. Randall MacIntosh, "Will Governor Act on Flint's Tainted Water? No Answer Yet," *FJ*, October 7, 1986.

27. David Waymire, "Pure Water Plan Unveiled by Blanchard," *FJ*, September 18, 1984.

28. Kaye Lafond and Rebecca Williams, "In 1994, Michigan OK'd Partial Pollution Cleanups. Now We Have 2,000 Contaminated Sites," Michigan Radio, April 17, 2018, https://www.michiganradio.org/post/1994-michigan-ok-d-partial-pollution -cleanups-now-we-have-2000-contaminated-sites.

29. Sarah Klein, "Reinventing Jim," *Metro Times*, July 24, 2002, https://www.metro times.com/detroit/reinventing-jim/Content?oid=2174175.

30. David Poulson, "Science Board Raises Concerns of Engler Critics," *FJ*, August 23, 1992.

31. Mike Stobbe, "Burned Out: Environmental Groups Losing Tired Leaders, Volun- teers, and Members," FJ, January 6, 1991.

32. All quotes from this section are from transcripts of city council meetings on Janu- ary 13, 2014; July 28, 2014; and May 17, 2017.

33. Adrian Hedden, "Councilman Leads Protest at Flint City Hall, Addresses Police Chases, Water Rates," *FJ*, July 14, 2014, https://www.mlive.com/news/flint/2014 /07/councilman_leads_protest_at_fl.html.

34. Curt Guyette, "In Flint, Michigan, Overpriced Water Is Causing People's Skin to Erupt in Rashes and Hair to Fall Out," *The Nation*, July 16, 2015, http://www .thenation.com/article/in-flint-michigan-overpriced-water-is-causing-peoples -skin-to-erupt-and-hair-to-fall-out/.

35. All quotes from city council meetings in this section are from transcripts of record- ings for January 13, 2014; July 28, 2014; and May 17, 2017.

36. Kristin Moore, "Statement from Mayor Weaver on Impending Tax Lien Notices Issued in Flint," posted by City of Flint, May 4, 2017, https://www.cityofflint.com /2017/05/04/statement-from-mayor-weaver-on-impending-tax-lien-notices-issued -in-flint/.

37. Concerned Pastors for Soc. Action v. Khouri, 217 F. Supp. 3d 960 (US District Court Eastern District of Michigan 2016)

38. Ron Fonger, "State Says Flint River Water Meets All Standards but More Than Twice the Hardness of Lake Water," *FJ*, May 23, 2014, https://www.mlive.com /news/flint/2014/05/state_says_flint_river_water_m.html.

39. FCC, transcript of public comment period, November 19, 2015.

40. USEPA FOIA, email from unidentified Flint resident to President Obama, Novem- ber 23, 2014.

41. USEPA FOIA, email from unidentified Flint resident to President Obama, Janu- ary 13, 2015; USEPA FOIA, letter to President Obama, January 13, 2015.

42. USEPA FOIA, email, September 20, 2015; USEPA FOIA, email, January 27, 2015.

43. USEPA FOIA, email, January 27, 2015.

44. USEPA FOIA EPA-R5-2016-003714, 9/20/2015 email.

45. USEPA FOIA, email from unidentified Flint resident to President Obama, to Thomas Poy, September 14, 2014.

46. Poy email.

47. USEPA FOIA, email to Jennifer Crooks, September 16, 2014.

48. USEPA FOIA, email from Crooks to Stephen Busch and Michael Prysby, September 17, 2014.

49. USEPA FOIA, email exchanges between Prysby to Jennifer Crooks and Stephen Busch, on September 17, 2014, September 17, 2014.

50. Prysby email.

51. Email exchanges between Prysby, Brent Wright, and Robert Bincsik.

52. USEPA FOIA, email from Michael Prysby to Jennifer Crooks, September 17, 2014.

53. USEPA FOIA, email to Jennifer Crooks, September 17, 2014.

54. USEPA FOIA, email from Jennifer Crooks to Flint resident, September 17, 2014.

55. USEPA FOIA, email from Jennifer Crooks to Thomas Poy, September 17, 2014.

56. Lockwood, Andrews and Newman, "Operational Evaluation Report City of Flint, Trihalomethane Formation Concern," November 2014, 1 and 4, https://www.greatlakeslaw.org/Flint/LAN_2014_Report.pdf.

57. City of Flint Public Notice, "Important Information about Your Drinking Water," January 2, 2015, https://www.cityofflint.com/wp-content/uploads/TTHM-Notification-Final.pdf.

58. City of Flint, "City of Flint 2014 Annual Water Quality Report."

59. USEPA FOIA, email from Jennifer Crooks to Mike Prysby, March 23, 2015.

60. USEPA FOIA, email from Jennifer Crooks to Mike Prysby, February 12, 2015.

61. USEPA FOIA EPA-R5-2016-003714, email from Jennifer Crooks to Mike Prysby, February 12, 2015.

62. USEPA FOIA, emails between Jennifer Crooks and Mike Prysby, March 25, 2015.

63. USEPA FOIA, email from Heather Shoven to several EPA officials, March 26, 2015.

64. US House of Representatives, *Hearing before the Committee on Oversight and Government Reform, Examining Federal Administration of the Safe Drinking Water Act in Flint, Michigan, Part II*, 114th Cong., 2nd sess., March 15, 2016, Serial No. 114–49, 231–32.

65. USEPA FOIA, FY 2014 Draft General ARDP, June 12, 2013, 17.

66. Draft General ARDP, 18.

67. USEPA FOIA, FY 2015 Draft General ARDP, September 17, 2014, 20.

68. USEPA FOIA, Michigan Community Water Supply Violation Report, June 2014, appendix A.

69. Michigan Community Water Supply Violation Report.

70. Michigan Community Water Supply Violation Report, appendix C.

71. William E. Ketchum III, "Water Advocates Walk from Detroit to Flint before Friday Protest at City Hall," Mlive.com, July 10, 2015, http://www.mlive.com/news/flint/index.ssf/2015/07/water_advocates_walk_from_detr.html.

72. Ron Fonger, "Flint Mayor Accepts Petitions but Not Call to End Use of Flint River," Mlive.com, August 31, 2015.

73. USEPA FOIA, email from unidentified Flint resident to President Obama, August 14, 2015.

74. USEPA FOIA, email from unidentified Flint resident to President Obama, September 15, 2015.

75. USEPA FOIA 03714, email to President Obama, May 15, 2015.

CHAPTER 4

1. Benjamin J. Pauli, *Flint Fights Back: Environmental Justice and Democracy in the Flint Water Crisis* (Cambridge, MA: MIT Press, 2019), 130.

2. Ron Fonger, "Flint Councilman Equates Water Troubles to 'Genocide' by Governor," *FJ*, April 6, 2015, https://www.mlive.com/news/flint/2015/04/flint_council man_claims_govern.html.

3. Pauli, *Flint Fights Back*, 130.

4. Flint Neighborhoods United website, accessed December 22, 2019, https://www .flintneighborhoodsunited.org/who-we-are-what-we-do/.

CHAPTER 5

1. Lindsey Smith, "Leaked Internal Memo Shows Federal Regulator's Concerns about Lead in Flint's Water," Michigan Radio, July 13, 2015, https://www.michigan radio.org/post/leaked-internal-memo-shows-federal-regulator-s-concerns-about -lead-flint-s-water.

2. Jim Lynch, "Whistle-Blower Del Toral Grew Tired of EPA 'Cesspool,'" *DN*, March 28, 2016, https://www.detroitnews.com/story/news/michigan/flint-water -crisis/2016/03/28/whistle-blower-del-toral-grew-tired-epa-cesspool/82365470/.

3. Donovan Hohn, "Flint's Water Crisis and the 'Troublemaker' Scientist," *New York Times*, August 16, 2016, https://www.nytimes.com/2016/08/21/magazine/flints -water-crisis-and-the-troublemaker-scientist.html.

4. Tiffany Stecker, Kevin Bogardus, and Sam Pearson, "Flint Crisis: EPA Official Warned of 'Public Health Disaster,'" E&E Daily, May 13, 2016, https://www.eenews .net/stories/1060037198.

5. Benjamin J. Pauli, *Flint Fights Back: Environmental Justice and Democracy in the Flint Water Crisis* (Cambridge, MA: MIT Press, 2019), 187–89.

6. Walters is the Flint resident who noticed her child's unexplained health problems and was the first person in Flint to relate the problem of lead contamination to the city's water. Walters's house was tested for lead in 2015, revealing extremely high levels of lead (104 µg/L in February, 397 µg/L in March, and 707 µg/L in April 2015). Additionally, a medical doctor found that one of her twins' blood lead level was 6.5 µg/dL, and an EPA inspection did not locate additional sources of lead in the home. Water in Walters's house was contaminated despite the fact that the water flowed into home through plastic pipes. M. A. Del Toral, USEPA Region 5, "High Lead Levels in Flint, Michigan—Interim Report," memo to Thomas Poy, June 24, 2015, http://flintwaterstudy.org/wp-content/uploads/2015/11/Miguels-Memo.pdf; CCE;

US House of Representatives Committee on Oversight and Government Reform, *Testimony of Lee-Anne Walters Examining Federal Administration of the Safe Drinking Water Act in Flint, Michigan*, 114th Cong. (Washington, DC: GPO, February 3, 2016).

7. Pauli, *Flint Fights Back.*
8. Kelsey J. Pieper, Rebekah Martin, Min Tang, Lee-Anne Walters, Jeffrey Parks, Siddhartha Roy, Christina Devine, and Marc A. Edwards, "Evaluating Water Lead Levels during the Flint Water Crisis," *Environmental Science and Technology* 52, no. 15 (2018): 8124–32.
9. Marc Edwards et al., "Lead Testing Results for Water Sampled by Residents," Flint Water Study, September 28, 2015, http://flintwaterstudy.org/information-for-flint -residents/results-for-citizen-testing-forlead-300-kits/.
10. Sara Jerome, "Feds Called in for Corrosion Control Treatment in Flint," Water Online, September 16, 2015, http://www.wateronline.com/doc/feds-called-in-for -corrosion-control-treatment-in-flint-0001.
11. Mona Hanna-Attisha, *What the Eyes Don't See: A Story of Crisis, Resistance, and Hope in an American City* (New York: One World, 2018).
12. Mona Hanna-Attisha, "Pediatric Lead Exposure in Flint, MI: Concerns from the Medical Community," Flint Water Studies Updates, September 24, 2015, http:// flintwaterstudy.org/wp-content/uploads/2015/09/Pediatric-Lead-Exposure-Flint -Water-092415.pdf.
13. "Test Update: Flint River Water 19X More Corrosive Than Detroit Water for Lead Solder; Now What?," Flint Water Study, September 11, 2015, http://flintwaterstudy .org/2015/09/test-update-flint-river-water-19x-more-corrosive-than-detroit-water -for-lead-solder-now-what/.
14. Email from Karen Tommasulo to Brad Wurfel," July 9, 2015, in M. Edwards, S. Roy, W. Rhoads, E. Garner, and R. Martin, "Chronological Compilation of e-mails from MDHHS Freedom of Information Act (FOIA)," request #2015-557 (2015).
15. Smith, "Leaked Internal Memo Shows Federal Regulator's Concerns about Lead in Flint's Water."
16. Email from Stephen Busch to Brad Wurfel, July 24, 2015, CCE.
17. Email from Brad Wurfel to Eric Brown, August 27, 2015, CCE.
18. Draft of news release, September 2, 2015, CCE.
19. Email from Brad Wurfel to Ron Fonger, September 9, 2015, CCE.
20. "Flint's Mayor Drinks Water from Tap to Prove It's Safe," WNEM Newsroom, July 9, 2015, https://www.wnem.com/news/flint-s-mayor-drinks-water-from-tap-to -prove-it/article_bad3b738–63f5–56d3-bcb7–52b40d591b20.html#ixzz30UM1z6BG.
21. Email from Ron Fonger to Brad Wurfel, September 11, 2015, CCE.
22. Email from Stephen Busch to Brad Wurfel, September 11, 2015, CCE.
23. Email from Liane Shekter-Smith to Stephen Busch et al., September 11, 2015, CCE.
24. Terry Gross, "Pediatrician Who Exposed Flint Water Crisis Shares Her 'Story of Resistance,'" National Public Radio, June 25, 2018, https://www.npr.org/sections /health-shots/2018/06/25/623126968/pediatrician-who-exposed-flint-water

-crisis-shares-her-story-of-resistance; Mona Hanna-Attisha, "Pediatric Lead Exposure in Flint, MI: Concerns from the Medical Community," Flint Water Studies Updates, September 24, 2015, http://flintwaterstudy.org/wp-content/uploads/2015/09/Pediatric-Lead-Exposure-Flint-Water-092415.pdf.

25. "Flint Mayor Dayne Walling Talks about Flint's Water Crisis, Emergency Managers, and the State Government," *Eclectablog*, October 29, 2015, http://www.eclectablog.com/2015/10/interview-flint-mayor-dayne-walling-talks-about-flints-water-crisis-emergency-managers-and-the-state-government.html.

26. John Wisely, "Snyder Announces $12 Million Plan to Fix Flint Water," *DFP*, October 8, 2015, http://www.freep.com/story/news/local/michigan/2015/10/08/snyder-flint-waterreconnect/73567778/.

27. Madeline Sturgeon, "True Grit: Pediatrician Proves Michigan Community's Water Was Poisoning Children," AAP News, November 11, 2015, http://www.aappublications.org/news/2015/11/11/Water111115.

28. Ron Fonger, "Karen Weaver Unseats Dayne Walling to Win Flint Mayor," *FJ*, November 4, 2015, https://www.mlive.com/news/flint/2015/11/karen_weaver_makes_history_ele.html.

29. Flint Water Advisory Task Force, "Final Report of the Flint Water Advisory Task Force," Office of the Governor, State of Michigan, March 2016, https://www.michigan.gov/documents/snyder/FWATF_FINAL_REPORT_21March2016_517805_7.pdf.

30. The pervasive malice or malign neglect of poor Black and Brown communities outlined in this book is a direct result of systemic racial bias, institutionalized power imbalances, and segregated residential patterns and resources. Since access to quality essential services within cities has been inextricably connected to the social standing or clout of community residents, people living in poor Black and Brown spaces have a long history of hosting the worst amenities and the most dangerous living conditions despite their efforts. While the tactics that promote malign neglect include attempts to ignore known deficits and mismanage resources within discarded spaces, the consequences of this level of disregard extend beyond property damage and long-term blight. Malign neglect perpetuates disadvantage in communities already gutted by decades of abandonment and disinvestment. It manifests as polluted community gardens and bankrupted schools with lead-tainted drinking fountains. It is targeted destruction, a form of bio-political warfare, laced with indifference for the people directly impacted.

31. Ron Fonger, "Flint Water Crisis Is Tidal Wave Issue in City's Mayoral Election," *FJ*, October 18, 2015, https://www.mlive.com/news/flint/2015/10/will_water.html.

32. FCC, transcript of public comment period, November 9, 2015.

33. FCC, transcript of public comment period. This is also the source of Shannon's quote below.

34. Steve Harmody, "Judge Sides with City of Flint in Court Fight over Water Rates," Michigan Radio, December 22, 2015, https://www.michiganradio.org/post/judge-sides-city-flint-court-fight-over-water-rates.

35. Ron Fonger, "Judge Signs Order for Lower Flint Water Rates, $15.7 Million Pay-back," *FJ*, August 17, 2015, https://www.mlive.com/news/flint/2015/08/judge_certifies_class_action_l.html.
36. Curt Guyette, "Lead Astray: An ACLU of Michigan Investigation Has Found a Stream of Irregularities in Flint's Water Tests," *Democratic Water*, September 14, 2015, http://www.aclumich.org/democracywatch/index.php/entry/lead-astray-an-aclu-of-michigan-investigation-has-found-a-stream-of-irregularities-in-flint-s-water-tests.
37. USEPA, letter from Miguel A. Del Toral, Regulations Manager, Ground Water and Drinking Water Branch, to Thomas Poy, June 24, 2015, 1–2, WG-15J, http://flintwaterstudy.org/wp-content/uploads/2015/11/Miguels-Memo.pdf.
38. Email from Mike Prysby to Pat Cook, April 24, 2015, CCE.
39. Email from Miguel Del Toral ("Troubling Development in Flint") sent to Thomas Poy, Jennifer Crooks, Nicholas Damato, and Rita Bair, April 22, 2015, made available through USEPA Region 5 FOIA request, Flint FOIA Production 5–120000044.
40. USEPA, "Optimal Corrosion Control Treatment Evaluation Technical Recommendations for Primacy Agencies and Public Water Systems," Office of Water (4606M), EPA 816-B-16-003, March 2016 (updated), https://www.epa.gov/sites/production/files/2019-07/documents/occtmarch2016updated.pdf; National Drinking Water Clearinghouse, "Tech Brief: Corrosion Control," February 1997, https://water-research.net/Waterlibrary/privatewell/corrosion.pdf.
41. "Drinking Water Lead and Copper Sampling Instructions from the City of Flint Water Plant," 2015, from FOIA request document EPA-R5-20 15–0112990000085.
42. Oliver Milman, "US Authorities Distorting Tests to Downplay Lead Content of Water," *The Guardian*, January 22, 2016.
43. Philadelphia Water Department, "Lead and Copper Rule Sampling Instructions," updated June 17, 2014, attachment in letter from Yanna Lambrinidou, PhD, to Philadelphia residents, January 23, 2016, http://flintwaterstudy.org/wp-content/uploads/2016/01/Dear-Philadelphia-Residents.pdf.
44. Email from Jennie Perey Saxe, PhD, environmental scientist, Drinking Water Branch, USEPA Region 3, July 16, 2008, http://flintwaterstudy.org/wp-content/uploads/2016/01/Dear-Philadelphia-Residents.pdf.
45. Email from Miguel Del Toral to Jennifer Crooks, Michael Prysby, et al., February 27, 2015, CCE.
46. USEPA, Lead Contamination Control Act, EPA 570 /9–89-AA, July 1989, https://www.epa.gov/sites/production/files/2015-09/documents/epalccapamphlet1989.pdf.
47. Richard L. Marcus, "Benign Neglect Reconsidered," *University of Pennsylvania Law Review* 148, no. 6 (2000): 2000–20043; B. Eichengreen, "From Benign Neglect to Malignant Preoccupation: US Balance of Payments Policy in the 1960s," *Economic Events, Ideas, and Policy: The 1960s and After*, ed. George Perry and James Tobin, 185–240 (Washington, DC: Brookings Institution Press, 2000).
48. Mary Doidge et al., "The Flint Fiscal Playbook: An Assessment of the Emergency

Manager Years (2011–2015)," MSU Extension White Paper, accessed November 18, 2019, https://www.canr.msu.edu/uploads/resources/pdfs/flint-fiscal-playbook.pdf; Eric Scorsone and Nicolette Bateson, *Long-Term Crisis and Systemic Failure: Taking the Fiscal Stress of America's Older Cities Seriously; Case Study: City of Flint, Michigan*, MSU Extension, September 2011, https://www.cityofflint.com/wp-content/uploads/Reports/MSUE_FlintStudy2011.pdf.

49. The t-statistic is a value that represents the magnitude and direction (e.g., positive or negative sign) of the difference in sample means. A p-value less than 0.05 indicates that the difference between the sample means is statistically significant.

50. Ron Fonger, "Michael Moore's Flint Water Movie Claims County Faked Kids' Lead Blood Tests," *FJ*, September 14, 2018, https://www.mlive.com/news/flint/2018/09/post_508.html; "Michael Moore's Fahrenheit 11/9—April Cook-Hawkins Film Clip," YouTube, accessed January 3, 2020, https://www.youtube.com/watch?reload=9&v=CtnANFLfH48.

51. Christopher Muller, Robert J. Sampson, and Alix S. Winter, "Environmental Inequality: The Social Causes and Consequences of Lead Exposure," *Annual Review of Sociology* 44 (July 2018): 263.

52. Robert J. Sampson and Alix Winter, "The Racial Ecology of Lead Poisoning: Toxic Inequality in Chicago Neighborhoods, 1995–2013," *DuBois Review: Social Science Research on Race* 13, no. 2 (2016): 261–83.

53. Richard Rabin, "The Lead Industry and Lead Water Pipes: 'A Modest Campaign,'" *AJPH* 98, no. 9 (September 2008): 1584–92.

54. Fayette F. Forbes, "A Very Brief Discussion of Lead Poisoning, Caused by Water Which Has Been Drawn through Lead Service Pipe," *Journal of the New England Water Works Association* 15, no. 1 (1900): 58–59.

55. Rabin, "Lead Industry and Lead Water Pipes."

56. Gerald Markowitz and David Rosner, "Cater to the Children: The Role of the Lead Industry in a Public Health Tragedy, 1900–1955," *AJPH* (January 2000): 36–46.

57. Markowitz and Rosner, 43.

58. US Congress, House of Representatives, *Lead Poisoning: Hearings before the Subcommittee on Health and the Environment of the Committee on Energy and Commerce*, 102nd Cong., 1st sess., including H.R. 2840, April 25 and July 26, 1991.

59. S. Klitzman, J. Caravanos, C. Belanoff, and L. Rothenberg, "A Multihazard, Multistrategy Approach to Home Remediation: Results of a Pilot Study," *Environmental Research*, 99, no. 3 (November 2005): 294–306.

60. M. Hauptman, R. Bruccoleri, and A. D. Woolf, "An Update on Childhood Lead Poisoning," *Clinical Pediatric Emergency Medicine* 18, no. 3 (2017): 181–92.

61. Jaime Raymond, William Wheeler, and Mary Jean Brown, "Inadequate and Unhealthy Housing, 2007 and 2009," *Morbidity and Mortality Weekly Report* 60, no. 1 (January 2011): 21–27; R. Raymond, M. Anderson, D. Feingold, D. Homa, and M. Brown, "Risk for Elevated Blood Lead Levels in 3- and 4-Year-Old Children," *Maternal and Child Health Journal* 13, no. 1 (2009): 40–47.

62. Rabin, "Lead Industry and Lead Water Pipes."

63. Ruth Ann Etzel and Sophie J. Balk, *Pediatric Environmental Health*, 4th ed. (Itasca, IL: American Academy of Pediatrics, 2018); Klitzman, Caravanos, Belanoff, and Rothenberg, "Multihazard, Multistrategy Approach to Home Remediation."

64. R. C. Sadler, J. LaChance, and M. Hanna-Attisha, "Social and Built Environmental Correlates of Predicted Blood Lead Levels in the Flint Water Crisis," *AJPH* (2017): 763–69.

65. Chammi P. Attanayake, Ganga M. Hettiarachchi, Ashley Harms, DeAnn Presley, Sabine Martin, and Gary M. Pierzynski, "Field Evaluations on Soil Plant Transfer of Lead from an Urban Garden Soil," *Journal of Environmental Quality* 43 (2014): 475–87.

66. A. Wood and B. J. Lence, "Assessment of Water Main Break Data for Asset Management," *Journal (American Water Works Association)* 98, no. 7 (July 2006): 76–86.

67. Arlene Karidis, "A Look at Brownfields: The Dirty, the Ugly, and the Potential for Turnaround," Waste Dive, January 21, 2016, https://www.wastedive.com/news/a -look-at-brownfields-the-dirty-the-ugly-and-the-potential-for-turnaroun/412478/.

68. Gwyneth Shaw, Beverly Ford, and Evelyn Larrubia, "EPA's 'Brownfields' Program Coming Up Short," Center for Public Integrity, September 25, 2012, updated May 19, 2014, https://publicintegrity.org/environment/epas-brownfields-program -coming-up-short/; Richard G. Opper, "The Brownfield Manifesto," *Urban Lawyer* 37, no. 1 (2005): 163–90; Anne Slaughter Andrew, "Brownfield Redevelopment: A State-Led Reform of Superfund Liability," *Natural Resources and Environment* 10, no. 3 (2006): 27–31.

69. Terence J. McGhee and E. W. Steel, *Water Supply and Sewerage*, 6th ed. (New York: McGraw-Hill Press, 2001).

70. Balvant Rajani, Yehuda Kleiner, and Jean-Eric Sink, "Exploration of the Relationship between Water Main Breaks and Temperature Covariates," *Urban Water Journal* 9, no. 2 (2012): 67.

71. N. E. Miller, "Winter Water-Main-Break Operations in Milwaukee," *Journal (American Water Works Association)* 68, no. 1 (January 1976): 10–11.

72. G. Baird, "Money Matters: Fasten Your Seat Belts: Main Breaks and the Issuance of Precautionary Boil-Water Notices," *Journal (American Water Works Association)*, 103, no. 3 (March 2011): 24–28; Bryon L. Livingston, Cliff Cate, Anna Pridmore, Jeffrey W. Heidrick, and Jim Geisbush, "Pipelines," 2016. *Out of Sight, Out of Mind, Not Out of Risk.* [N.p.], American Society of Civil Engineers.

73. K. C. Thompson, J. Gray, and Ulrich Borchers, *Water Contamination Emergencies: Managing the Threats* (Cambridge: Royal Society of Chemistry, 2013).

74. Martin Allen, Robert Clark, Joseph A. Cotruvo, and Neil Grigg, "Drinking Water and Public Health in an Era of Aging Distribution Infrastructure," *Public Works Management and Policy* 23, no. 4 (2018): 301; Lori E. A. Bradford et al., "There Is No Publicity Like Word of Mouth . . . Lessons for Communicating Drinking Water Risks in the Urban Setting," *Sustainable Cities and Society* 29 (February 2017): 23–40.

75. Victoria Morckel, "Why the Flint, Michigan, USA Water Crisis Is an Urban Planning Failure," *Cities* 62 (2017): 23–27.

76. Roberto Acosta, "Water Main Break Leads to Traffic Detour in Downtown Flint," *FJ*, October 20, 2019, https://www.mlive.com/news/flint/2019/10/water-main -break-leads-to-traffic-detour-in-downtown-flint.html; Zahra Ahmad, "Water Main Break Cancels Flint City Council Meetings," *FJ*, October 28, 2019, https:// www.msn.com/en-us/sports/more-sports/water-main-break-cancels-flint-city -council-meetings/ar-AAJuZH1; "Main Break Sparks Boil Filtered Water Advisory in Flint," *DN*, December 25, 2017, https://amp.detroitnews.com/amp/108918514.

77. Lindsey Smith, "Boil Water Advisory Adds to Confusion over How to Make Flint Tap Water Safe to Drink," Michigan Radio, February 11, 2016, https://www .michiganradio.org/post/boil-water-advisory-adds-confusion-over-how-make -flint-tap-water-safe-drink.

78. Kris Maher, "Lead Levels in Flint Remain Extremely High at Some Locations; Children, Pregnant Mothers Should Avoid Untested Water, Officials Warn," *Wall Street Journal* (online), February 1, 2016.

79. Michigan Department of Environmental Quality FOIA 6797–15, "City of Flint, Water Treatment Plant Sanitary Survey," April 9, 2013.

80. USEPA, "Permeation and Leaching," Office of Ground Water and Drinking Water Distribution System Issue Paper, August 15, 2002, https://www.epa.gov/sites /production/files/2015–09/documents/permeationandleaching.pdf.

81. Edward C. Glaza and Jae K. Park, "Permeation of Organic Contaminants through Gasketed Pipe Joints," *Journal (American Water Works Association)* 84, no. 7 (1992): 92–100; Thomas M. Holsen et al., "The Effect of Soils on the Permeation of Plastic Pipes by Organic Chemicals," *Journal (American Water Works Association)* 83, no. 11 (1991): 85–91.

82. USEPA, "Permeation and Leaching."

83. P. J. Landrigan, C. B. Schechter, J. M. Lipton, M. C. Fahs, and J. Schwartz, "Environmental Pollutants and Disease in American Children: Estimates of Morbidity, Mortality, and Costs for Lead Poisoning, Asthma, Cancer, and Developmental Disabilities," *Environmental Health Perspectives* 110, no. 7 (2002): 721–28; H. L. Needleman et al., "The Long-Term Effects of Exposure to Low Doses of Lead in Childhood—An 11-Year Follow-up Report," *New England Journal of Medicine* 322, no. 2 (January 1990): 83–88; H. L. Needleman et al., "Bone Lead Levels and Delinquent Behavior," *Journal of the American Medical Association* 275, no. 5 (February 1996): 363–69.

84. Advisory Committee on Childhood Lead Poisoning Prevention, "Recommendations for Blood Lead Screening of Young Children Enrolled in Medicaid: Targeting a Group at High Risk," CDC, December 8, 2000, https://www.cdc.gov/mmwr /preview/mmwrhtml/rr4914a1.htm.

85. Council On Environmental Health, "Prevention of Childhood Lead Toxicity," *Pediatrics* 38, no. 1 (2016): e20161493.

86. A. R. Kemper, L. M. Cohn, K. E. Fant, K. J. Dombkowski, and S. R. Hudson, "Follow-up Testing among Children with Elevated Screening Blood Lead Levels," *Journal of the American Medical Association* 293, no. 18 (2005): 2232–37.

87. J. Dickman and Safer Chemicals, Healthy Families, *Children at Risk: Gaps in State Lead Screening Policies*, SaferChemicals.org, January 2017, https://saferchemicals .org/wp-content/uploads/2017/01/saferchemicals.org_children-at-risk-report.pdf.
88. "Learn about Lead Testing in Massachusetts and What a Result Means for Your Child," Mass.gov., accessed October 18, 2020, https://www.mass.gov/service -details/learn-about-lead-testing-in-massachusetts-and-what-a-result-means-for -your-child.
89. MDHHS, *Medicaid Provider Manual*, 2018, accessed August 1, 2018, http://www .mdch.state.mi.us/dch-medicaid/manuals/MedicaidProviderManual.pdf.
90. GAO, *Lead Poisoning: Federal Health Care Programs Are Not Effectively Reaching At-Risk Children*, GAO-HEHS-99–18, US General Accounting Office, January 1999, https://www.gao.gov/archive/1999/he99018.pdf.
91. California Department of Health Services, *Guidance Manual for Implementing Fingerstick Sampling* (Emeryville, CA: Childhood Lead Poisoning Prevention Branch, California Department of Health Services, September 1997).
92. MDHHS, *Medicaid Provider Manual*, 2019, accessed October 18, 2020, http://www .mdch.state.mi.us/dch-medicaid/manuals/MedicaidProviderManual.pdf; MDHHS, *Medicaid Provider Manual*, 2018; MDHHS, "Blood Lead Level Quick Reference for Primary Care Providers," 2015, accessed December 5, 2019, https://www.michigan .gov/documents/deq/ProviderQuickReference_Sept2015_501831_7.pdf.

CHAPTER 6

1. Transcript of Michigan State of the State Address by Governor Rick Snyder, January 19, 2016, https://www.c-span.org/video/?403297-1/michigan-governor-rick -snyder-state-state-address.
2. Throughout the chapter, quotations from the 2016 congressional hearings on the Flint water crisis are from transcripts of testimonies at the February 3, February 10, and March 15, 2016 hearings, student transcription, accessed October 18, 2020, https://libguides.umflint.edu/watercrisis/commentary.
3. Ron Fonger, "Genesee County Health Officer Retires, National Search Begins for Replacement," *FJ*, October 5, 2017, https://www.mlive.com/news/flint/2017/10 /genesee_county_will_try_nation.html.
4. Ron Fonger, "Public Never Told, but Investigators Suspected Flint River Tie to Legionnaires' in 2014," *FJ*, January 16, 2016, https://www.mlive.com/news/flint /2016/01/documents_show_agencies_knew_o.html.
5. Yahoo News Video, "Erin Brockovich Calls the Flint Hearings 'A Waste of Everybody's Time,'" March 17, 2016, https://finance.yahoo.com/video/erin-brockovich -calls-flint-hearings-184420353.html.
6. Liam Stack, "Michigan Gave State Employees Purified Water as It Denied Crisis, Emails Show," *New York Times*, January 29, 2016, https://www.nytimes.com/2016 /01/30/us/flint-michigan-purified-water.html.
7. Dominic Adams, "Here's How Flint Went from Boom Town to Nation's Highest Poverty Rate," *FJ*, September 21, 2017, https://www.mlive.com/news/flint/2017/09 /heres_how_flint_went_from_boom.html.

8. Gary Ridley, "President Obama to Visit Flint Next Week for Firsthand Look at Water Crisis," *FJ*, April 27, 2016, https://www.mlive.com/news/flint/2016/04/president_obama_to_visit_flint.html.

9. Paul Egan and Kat Stafford, "Obama to Flint: 'Don't Lose Hope,' Drinks the Water," *DFP*, May 4, 2016, https://www.freep.com/story/news/local/michigan/flint-water-crisis/2016/05/04/obama-lands-flint-water-crisis-visit/83915846/.

10. President Barack Obama, transcript of speech in Flint on the Water Crisis, May 4, 2016, https://time.com/4318909/barack-obama-speech-flint-michigan-transcript/.

11. "Michael Moore on Pres. Obama: He 'Might as Well Stay Home,'" CNN.com, 2016, accessed December 18, 2019, https://www.cnn.com/videos/tv/2016/05/04/michael-moore-on-flint-crisis-lead-live.cnn.

12. Francesca Chambers, "'The Man Was Just Thirsty!' White House Insists Obama Didn't Drink Flint Water the Second Time as a 'Stunt'—and Says Reporters Made Him Do It the First Time (and Yes He May Have Ate Lead Paint Chips as a Child)," DailyMail.com, May 5, 2016, https://www.dailymail.co.uk/news/article-3575655/The-man-just-thirsty-White-House-insists-Obama-didn-t-drink-Flint-water-second-time-stunt-says-reporters-time-yes-ate-lead-paint-chips-child.html.

13. Max Plenke, "22 Powerful Photos Show the Devastating Reality of Flint's Water Crisis," Yahoo News, January 21, 2016, https://www.yahoo.com/news/22-powerful-photos-show-devastating-134300579.html.

14. Ed Henry, "Obama Takes Plunge, Swims in Gulf," CNN.com, August 14, 2010, https://www.cnn.com/2010/POLITICS/08/14/obama.gulf.swim/index.html; Suzanne Goldenberg, "BP Oil Spill: Barack Obama Dives into Safety Debate with Gulf of Mexico Swim," *The Guardian*, August 15, 2010, https://www.theguardian.com/environment/2010/aug/15/barack-obama-swim-gulf-florida.

15. C-SPAN, "President Obama Remarks in Flint, Michigan," C-SPAN.org, May 4, 2016, https://www.c-span.org/video/?409040-1/president-obama-addresses-residents-flint-michigan&start=1207.

16. "Michael Moore on Pres. Obama: He 'Might as Well Stay Home.'"

17. Interview with Flint resident, May 27, 2017.

18. Quotes from emails and letters to President Obama are from USEPA FOIA; correspondence was sent on January 13, 2015; January 27, 2015; and February 10, 2015.

19. Letter by Tinka G. Hyde to concerned resident, May 16, 2015, WG-15J, from DEQ FOIA request 6797–15.

20. This was highlighted to me by Ward 1 councilman Eric Mays.

21. Amanda Emery, "Giggles the Pig Withdraws from Flint Mayoral Race, According to Facebook Post," *FJ*, June 10, 2015, https://www.mlive.com/news/flint/2015/06/giggles_the_pig_withdraws_from.html; David Lohr, "Giggles the Pig Runs for Mayor against 2 Convicted Felons," *Huffington Post*, May 16, 2015, https://www.huffpost.com/entry/giggles-the-pig_n_7293404.

22. Steve Carmody, "Defamation Lawsuit Filed Involving Key Flint Water Crisis Figures," Michigan Radio, July 23, 2018, https://www.michiganradio.org/post/defamation-lawsuit-filed-involving-key-flint-water-crisis-figures; Steve Carmody, "Judge Dismisses Defamation Lawsuit Involving Flint Water Crisis Figures,"

Michigan Radio, March 21, 2019, https://www.michiganradio.org/post/judge
-dismisses-defamation-lawsuit-involving-flint-water-crisis-figures.

23. Benjamin J. Pauli, *Flint Fights Back: Environmental Justice and Democracy in the Flint Water Crisis* (Cambridge, MA: MIT Press, 2019), 219.

24. Letter by water activists was posted on May 10, 2018, titled "Complaint Request Letter," Flint Complaints.com, http://www.flintcomplaints.com/.

25. Edwards v. Schwartz et al., District Court of Western Virginia, Case No. 7:18-cv-00378-MFU: 4 (W.D. Va. Mar. 19, 2019).

26. Carmody, "Judge Dismisses Defamation Lawsuit Involving Flint Water Crisis Figures."

27. Michigan Attorney General's Office, "Schuette Charges MDHHS Director Lyon, Four Others with Involuntary Manslaughter in Flint Water Crisis," Michigan.gov, June 14, 2017, https://www.michigan.gov/ag/0,4534,7–359–82917_78314 –423854—,00.html.

28. Complaint, People v. Lyons, No. 16–0003 (Mich. 67th D. Ct., June 14, 2017): 1, https://www.michigan.gov/documents/ag/Nick+Lyon+Warrant+Packet_575706 _7.pdf.

29. People v. Lyons, 2.

30. People v. Lyons.

31. People v. Lyons, 4.

32. People v. Lyons, 6.

33. People v. Lyons, 6–8.

34. People v. Lyons, 10.

35. AP, "Michigan's Ex-Drinking Water Regulator Takes Deal in Flint Water Investigation," *DFP*, January 7, 2019, https://www.freep.com/story/news/local/michigan /2019/01/07/ex-water-regulator-takes-deal-flint-water-investigation/2503990002/.

36. AP, "Michigan's Ex-Drinking Water Regulator Takes Deal in Flint Water Investigation"; Paul Egan, "2 DEQ Officials Plead No Contest in Flint Water Crisis Case," *DFP*, December 26, 2018, https://www.freep.com/story/news/local/michigan /flint-water-crisis/2018/12/26/flint-water- Dec. 26, 2018crisis-2018/2414377002/.

37. Ron Fonger, "Flint Water Plant Manager Says He Put State on Notice with Bad Info on Lead," Mlive.com, March 22, 2018, https://www.mlive.com/news/flint /index.ssf/2018/03/flint_water_plant_manager_says.html.

38. Egan, "2 DEQ Officials Plead No Contest in Flint Water Crisis Case."

39. Chad Livengood, "AG's Office Drops Criminal Charges in Flint Water Cases, Starts New Investigation," *Craine's Detroit Business*, June 13, 2019, https://www .crainsdetroit.com/government/ags-office-drops-criminal-charges-flint-water -cases-starts-new-investigation.

40. Ron Fonger, "Flint Lawmakers Push to Change Statute of Limitations for Flint Water Investigation," *FJ*, January 16, 2020, https://www.mlive.com/news/flint /2020/01/lawmakers-push-to-change-statute-of-limitations-for-flint-water -investigation.html.

41. "Flint Legislators Seek to Increase Opportunity for Justice in Water Crisis,"

SenateDems.com, August 28, 2019, https://senatedems.com/ananich/2019/08/28
/flint-legislators-seek-to-increase-opportunity-for-justice-in-water-crisis/.

42. Fonger, "Flint Lawmakers Push to Change Statute of Limitations for Flint Water
Investigation."

43. Mitch Smith, "Flint Water Prosecutors Drop Criminal Charges, with Plans to
Keep Investigating," *New York Times*, June 13, 2019, https://www.nytimes.com/2019
/06/13/us/flint-water-crisis-charges-dropped.html.

44. Status Conference: RE Flint Water Cases, Case No. 16–10444, July 26, 2017: 11.
United States District Court Eastern District of Michigan, Judge Judith E. Levy,
presiding.

45. Status Conference: RE Flint Water Cases, 16.

46. Status Conference: RE Flint Water Cases, 66.

47. Co-liaison Opposition Brief to Motion for Replacement of Co-liaison Counsel
and Cross-Motion to Discharge Interim Co-lead Class Counsel, 1–2, Case No
5:16-cv-10444-JEL (MKM consolidated), Judge Judith E. Levy, presiding, filed
April 9, 2018.

48. Co-liaison Opposition Brief to Motion for Replacement of Co-liaison Counsel and
Cross-Motion to Discharge Interim Co-lead Class Counsel, 4–5.

49. Co-liaison Opposition Brief to Motion for Replacement of Co-liaison Counsel and
Cross-Motion to Discharge Interim Co-lead Class Counsel, 6.

50. Veolia Water North Operating Services, *Corrected Proposal Re Preliminary Discovery Plan*, 8, Civil Action No 5:16-cv-10444-JEL (MKM consolidated), Judge
Judith E. Levy presiding, filed October 9, 2017.

51. Status Conference: RE Flint Water Cases, Case No. 16–10444, September 12,
2018: 50.

52. Wald v. Snyder, First Amended Class Action Complaint, September 12, 2016, 4,
Eastern District of Michigan, mied-5:2016-cv-10444.

53. Status Conference: RE Flint Water Cases, Case No. 16–10444. September 26,
2018: 32.

54. Status Conference: RE Flint Water Cases, 33–34.

55. Status Conference: RE Flint Water Cases, 43.

56. Guertin v. Michigan, no. 17-1769 (6th Cir. 2019), p. 50; complaint filed June 27, 2016.

57. Gaddy et al. v. Flint et al., First Complaint, 184–85, Eastern District of Michigan,
5:2017-cv-11166.

58. Wald v. Snyder, First Amended Class Action Complaint, September 12, 2016, 68.

59. United States District Court, Eastern District of Michigan, Southern Division, In
Re Flint Water Cases, Case No. 16–10444, Status Conference, Judge Judith E. Levy
presiding, February 6, 2019, 27, http://www.mied.uscourts.gov/PDFFIles/FWC
-2019_02_06_Status_Conference_Transcript.pdf.

60. Jan Worth-Nelson, "Water Class Action Attorneys Detail Progress of Civil Cases;
Not Everyone Satisfied," *East Village Magazine*, June 28, 2019, https://www.east
villagemagazine.org/2019/06/28/water-class-action-attorneys-detailing-progress
-of-civil-cases-not-everyone-satisfied/.

61. Worth-Nelson, "Water Class Action Attorneys Detail Progress of Civil Cases; Not Everyone Satisfied."

62. State of Michigan, "Terms of Settlement of Flint Water Cases against the State of Michigan," accessed October 1, 2020, https://www.michigan.gov/documents/ag /Terms_of_Settlement_699810_7.pdf.

63. Worth-Nelson, "Water Class Action Attorneys Detail Progress of Civil Cases; Not Everyone Satisfied."

CHAPTER 7

1. "Dr. Karen Weaver Swearing In," YouTube, November 9, 2015, https://www .youtube.com/watch?v=yUmvYTBQf8s; "Karen Weaver Takes Oath as Flint's First Woman Mayor," *DFP*, November 9, 2015, https://www.freep.com/story/news/local /michigan/2015/11/09/flint-mayor-karen-weaver-first-black-flint-mayor/75481182/.

2. "Marion Coates Williams," Karen about Flint, March 11, 2015, http://karenabout flint.com/marion-coates-williams/; "Negro Wins Flint School Board Post," *Jet*, May 2, 1963, vol. 24, no. 2, p. 45, accessed December 19, 2019, https://books.google .com/books?id=zMIDAAAAMBAJ&pg=PA45&lpg=PA45&dq=Dr.+T.+Wendell +Williams+board+of+education+election+flint&source=bl&ots=bFESvoHMC -&sig=ACfU3U1_m33uftZHvR2bmdOU1cth_huCPw&hl=en&ppis=_e&sa=X &ved=2ahUKEwiA4cWj2_HmAhVRZN8KHcqLB5UQ6AEwAXoECAwQAQ #v=onepage&q=Dr.%20T.%20Wendell%20Williams%20board%20of%20education %20election%20flint&f=false.

3. "Dr. Karen Weaver Swearing In."

4. "Dr. Karen Weaver Swearing In."

5. City of Flint, "Receiver ship Transition Advisory Board (RTAB)," CityofFlint.com, accessed November 1, 2019, https://www.cityofflint.com/rtab/.

6. Ron Fonger, "Mayor Karen Weaver Declares Water Crisis State of Emergency in Flint," *FJ*, January 19, 2019, https://www.mlive.com/news/flint/2015/12/mayor _karen_weaver_declares_st.html.

7. Concerned Pastors for Social Action v. Khouri, No. 16–2628 (6th Cir. 2016), 7.

8. Concerned Pastors for Social Action v. Khouri.

9. City of Flint, "CORE: What Is the CORE Program About?," CityofFlint.com, accessed December 13, 2019, https://www.cityofflint.com/wp-content/uploads/FAQ. pdf; Ron Fonger, "Michigan DEQ Says CORE Has Visited Flint Residents in All but 140 Homes," *FJ*, March 15, 2018, https://www.mlive.com/news/flint/2018/03 /state_deq_says_core_has_visite.html.

10. Letter to Keith Creagh and Mayor Karen Weaver from the USEPA, March 17, 2017, epa.gov, https://www.epa.gov/sites/production/files/2017–03/documents /signedsrfletter31717.pdf.

11. Paul Egan, "$12M in Legal Bills for Flint—$3.4M for Snyder," *DFP*, April 5, 2017, https://www.freep.com/story/news/local/michigan/flint-water-crisis/2017/04/05 /flint-michigan-legal-fees/100066936/.

12. "Mayor Karen Weaver: Two More Years of Water Filter Use for Flint," *Eclectablog*,

March 13, 2017, http://www.eclectablog.com/2017/03/mayor-karen-weaver-two
-more-years-of-water-filter-use-for-flint.html.

13. "Number of Service Lines That Need Replacing in Flint Rises to 29,100, According
to Study," posted by Kristin Moore, CityofFlint.com, December 1, 2016, https://
www.cityofflint.com/2016/12/01/number-of-service-lines-that-need-replacing-in
-flint-rises-to-29100-according-to-study/.

14. Ryan Felton, "Fixing Flint's Contaminated Water System Could Cost $216M, Re-
port Says," *The Guardian*, June 6, 2016, https://www.theguardian.com/us-news
/2016/jun/06/flint-water-crisis-lead-pipes-infrastructure-cost.

15. Alex Kellogg, "Flint Mayor Calls for Immediate Removal of Corroded Lead Pipes,"
The Guardian, February 2, 2016, https://www.theguardian.com/us-news/2016
/feb/02/flint-water-crisis-lead-pipes-replaced.

16. Tracy Connor, "Michigan Gov. Rick Snyder Has No Plan to Remove Poison Pipes
in Flint: Flint Residents Sound-Off on Water Crisis at MSNBC Town Hall," *NBC
News*, January 28, 2016, https://www.nbcnews.com/storyline/flint-water-crisis
/michigan-gov-rick-snyder-has-no-plan-remove-poison-pipes-n505376.

17. Gary Ridley, "Flint Water Crisis Leaves City Finances in 'Very Precarious Situation,'" dated January 26, 2016, on Mlive.com., https://www.mlive.com/news
/flint/2016/01/flint_water_crisis_leaves_city.html.

18. "Video: Flint Residents Protest Paying for Poisoned Water," wsws.org, January 26,
2016, https://www.wsws.org/en/articles/2016/01/26/flin-j26.html.

19. Concerned Pastors for Social Action v. Khouri.

20. Settlement Agreement, Concerned Pastors for Social Action v. Khouri, No. 16–
2628 (6th Cir. 2016), March 21, 2017, https://www.clearinghouse.net/chDocs
/public/PB-MI-0011–0007.pdf.

21. Merritt Kennedy, "Judge Approves $97 Million Settlement to Replace Flint's
Water Lines," National Public Radio, March 28, 2017, https://www.npr.org
/sections/thetwo-way/2017/03/28/521786192/judge-approves-97-million-settlement
-to-replace-flints-water-lines.

22. Oona Goodin-Smith, "State to Start Closing Some Flint Water Distribution Sites,"
Mlive.com, July 26, 2017, http://www.mlive.com/news/flint/index.ssf/2017/07
/flint_water_distribution_sites_1.html; "Water Sites," CityofFlint.com, accessed
December 30, 2017 https://www.cityofflint.com/water-sites/.

23. AP, "Flint City Administrator's Whistleblower Lawsuit Reinstated by Appeals
Court," Michigan Radio, September 20, 2018, https://www.michiganradio.org
/post/flint-city-administrators-whistleblower-lawsuit-reinstated-appeals-court.

24. Emma Winowiecki, "Judge Dismisses Former Flint Official's Lawsuit against City,
Mayor Weaver," Michigan Radio, August 10, 2017, https://www.michiganradio.org
/post/judge-dismisses-former-flint-officials-lawsuit-against-city-mayor-weaver.

25. Steve Carmody and Doug Tribou, "Jury Sides with City of Flint in Whistleblower
Lawsuit," Michigan Radio, May 14, 2019, https://www.michiganradio.org/post
/jury-sides-city-flint-whistleblower-lawsuit.

26. FCC, transcript of public comment period, May 18, 2017. This is the source of the

residents' quotes in the next few paragraphs and later in the chapter where the May 2017 council meeting is mentioned.

27. Steve Carmody, "Board Approves Recall Petition Language against Flint Mayor Karen Weaver," Michigan Radio, March 8, 2017, https://www.michiganradio.org /post/board-approves-recall-petition-language-against-flint-mayor-karen-weaver.

28. Carmody, "Board Approves Recall Petition Language against Flint Mayor Karen Weaver."

29. FCC, transcript of public comment period, September 11, 2015.

30. Oona Goodin-Smith, "Flint Mayor Karen Weaver Survives Recall Vote with Landslide Victory," FJ, November 8, 2017, https://www.mlive.com/news/flint/2017/11 /flint_mayor_karen_weaver_wins.html.

31. Mitch Smith, "Flint Mayor, Ushered In to Fix Water Crisis, Now Faces Recall," New York Times, November 6, 2017, https://www.nytimes.com/2017/11/06/us /flint-mayor-karen-weaver-recall-water.html.

32. Hughey Newsome, "Open Letter to Flint City Council," FlintBeat.com, March 2019, http://flintbeat.com/wp-content/uploads/2019/03/RESIG-OPEN-LETTER -final-clean-copy.pdf.

33. Newsome, 1.

34. Ron Hilliard, "Flint City Council Claims City Staffers Boycotting Meetings," NBC25News.com, September 10, 2019, https://nbc25news.com/news/local /flint-city-council-claims-city-staff-boycotting-meetings.

35. Spectacle Productions, "Flint City Council Meeting Erupts," Facebook, September 4, 2019, https://www.facebook.com/watch/?v=433247167398189.

36. Spectacle Productions, "Flint City Council Meeting Erupts."

37. Clarissa Hamlin, "Four Years Later in Flint: Residents Rally over Shut Down in Free Bottled Water Program," NewsOne.com, April 11, 2018, https://newsone. com/3792133/flint-water-crisis-2018-residents-rally-bottled-service-ends-update/.

38. Nicole Chavez, "Michigan Will End Flint's Free Bottled Water Program," CNN, April 7, 2018, https://www.cnn.com/2018/04/07/us/flint-michigan-water-bottle -program-ends/index.html.

39. Kelsey J. Pieper, Rebekah Martin, Min Tang, Lee-Anne Walters, Jeffrey Parks, Siddhartha Roy, Christina Devine, and Marc A. Edwards, "Evaluating Water Lead Levels during the Flint Water Crisis," Environmental Science and Technology 52, no. 15 (2018): 8124–32.

40. Terese M. Olson, Madeleine Wax, James Yonts, Keith Heidecorn, Sarah-Jane Haig, David Yeoman, Zachary Hayes, Lutgarde Raskin, and Brian R. Ellis, "Forensic Estimates of Lead Release from Lead Service Lines during the Water Crisis in Flint, Michigan," Environmental Science and Technology Letters 4, no. 9 (2017): 356–61; Ahmed A. Abokifa and Pratim Biswas, "Modeling Soluble and Particulate Lead Release into Drinking Water from Full and Partially Replaced Lead Service Lines," Environmental Science and Technology 51, no. 6 (2017): 3318–26.

41. Chris Mills Rodrigo, "EPA Head Says Water in Flint 'Safe to Drink,'" The Hill, June 3, 2019, https://thehill.com/policy/energy-environment/446723-epa-head -water-in-flint-meets-quality-standards.

42. Ron Fonger, "Flint Mayor Disputes EPA Official's Claim That Water Is Safe," *FJ*, June 5, 2019, https://www.mlive.com/news/flint/2019/06/flint-mayor-disputes-epa-officials-claim-that-water-is-safe.html.

43. FCC, transcript of public comment period, November 29, 2018.

44. FCC, transcript of public comment period.

45. City of Flint, FAST Start Pipe Replacement Program website, accessed February 16, 2019, https://www.cityofflint.com/fast-start/.

46. AP Archive, "First Lead Water Pipes Replaced in Flint, US," YouTube, March 5, 2016, https://www.youtube.com/watch?v=LJ9CdKs-ZyM.

47. Harold C. Ford, "Pipe Replacement Crews Dig In: 'This Is Personal,'" *East Village Magazine*, July 29, 2017, https://www.eastvillagemagazine.org/2017/07/29/pipe-replacement-crews-dig-in-this-is-personal/.

48. Kristin Moore, "Mayor Weaver Holds Forum to Update Residents on Next Phase of FAST Start, Other Water Recovery Efforts," posted February 2, 2018, https://www.cityofflint.com/2018/02/02/mayor-weaver-holds-forum-to-update-residents-on-next-phase-of-fast-start-other-water-recovery-efforts/.

49. Kristin Moore, "Mayor Weaver Holds Forum to Update Residents on Next Phase of FAST Start, Other Water Recovery Effort," CityofFlint.com, February 2, 2018, https://www.cityofflint.com/2018/02/02/mayor-weaver-holds-forum-to-update-residents-on-next-phase-of-fast-start-other-water-recovery-efforts/.

50. Presentation by Ed Thayer, AECOM project manager, FCC meeting, July 23, 2018.

51. Declaration of Dr. Eric Schwartz, Concerned Pastors for Soc. Action v. Khouri, 217 F. Supp. 3d 960 (E.D. Mich. 2016), 5, filed October 1, 2018.

52. Declaration of Dr. Eric Schwartz, 8–9.

53. Madeline Ciak and Joel Feick, "Mayor Announces That All High Priority Lead, Iron Pipes Have Been Replaced in Flint," NBC25News.com, December 4, 2018, https://nbc25news.com/news/flint-water-woes/report-mayor-expected-to-announce-that-all-lead-iron-pipes-have-been-replaced-in-flint.

54. City of Flint, "Goyette Mechanical Water Service Line Replacements Zones 2, 3, 6, 8," Department of Public Works, Contract 17-010, signed March 23, 2017, https://assets.nrdc.org/sites/default/files/contract-goyette-flint-20170502.pdf?_ga=2.29308540.1996221421.1600915422-124799177.1600915422.

55. Zahra Ahmad, "Flint Mayor Has Raised More Than $250,000 for Next Election," *FJ*, January 11, 2019, https://www.mlive.com/news/flint/2019/01/flint-mayor-has-raised-more-than-250000-for-next-election.html.

56. "Many of the Homes Targeted by Flint's FAST Start Did Not Have High Lead Levels," NBC25News.com, July 14, 2016, https://nbc25news.com/news/local/many-of-the-homes-targeted-by-flints-fast-start-did-not-have-high-lead-levels.

57. Letter from Eric Oswald, Michigan DEQ, "Lead Service Lead Removal Activities," to Mayor Karen Weaver, May 31, 2018, Michigan.gov, https://www.michigan.gov/flintwater/0,6092,7-345-76292_76364-376646--,00.html.

58. Zahra Ahmad, "Flint City Council Approves Additional $1.1 Million to AECOM Contract despite Pleas against It," Mlive.com, December 11, 2018, https://www.mlive.com/news/flint/2018/12/

flint-city-council-approves-additional-11-million-to-aecom-contract-despite-pleas
-against-it.html.

59. Ahmad, "Flint City Council Approves Additional $1.1 Million to AECOM Con-
tract despite Pleas against It."

60. Zahra Ahmad, "Flint to Use Predictive Model to Plan for Phase 6 of Service Line
Replacements," Mlive.com, December 6, 2018, https://www.mlive.com/news
/2018/12/flint-to-use-predictive-model-to-plan-for-phase-6-of-service-line
-replacements.html.

61. Ahmad, "Flint to Use Predictive Model to Plan for Phase 6 of Service Line
Replacements."

62. Motion to Enforce Settlement Agreement, Concerned Pastors for Soc. Action v.
Khouri, 217 F. Supp. 3d 960 (E.D. Mich. 2016), 34, filed August 21, 2018.

63. Ahmad, "Flint City Council Approves Additional $1.1 Million to AECOM Con-
tract despite Pleas against It"; David Eggert, "Michigan Sues Flint after Council
Refuses to OK Water Deal," *Chicago Tribune*, June 28, 2017, https://www.chicago
tribune.com/news/nationworld/midwest/ct-flint-water-michigan-lawsuit
-20170628-story.html; Dominic Adams, "Flint Not Fulfilling Water Crisis Recovery
Obligations, Attorneys Claim," Mlive.com, July 24, 2018, https://www.mlive.com
/news/flint/2018/07/flint_not_obeying_federal_sett.html.

64. Steve Carmody, "Flint and MDEQ Locked in Dispute over City's Water System,"
Michigan Radio, June 12, 2018, https://www.michiganradio.org/post/flint-and
-mdeq-locked-dispute-over-citys-water-system; Drew Moore and Jason Lorenz,
"MDEQ: Flint Is Again in Violation of Safe Drinking Water Act," NBC25News.com,
June 4, 2018, https://nbc25news.com/news/local/mdeq-flint-is-again-in-violation
-of-safe-water-drinking-act.

65. Jim Malewitz, "Flint Finds Replacing Lead Pipes Isn't Easy. Even When State Promises to Pay," Bridge, November 28, 2018, https://www.bridgemi.com/michigan
-environment-watch/flint-finds-replacing-lead-pipes-isnt-easy-even-when-state
-promises-pay; Eggert, "Michigan Sues Flint after Council Refuses to OK Water
Deal."

66. City Council Meeting Transcript. November 27, 2017.

67. FCC meeting, transcript of public comment period, September 5, 2018. This is the
source of Parks's quotes in the next few paragraphs.

68. FCC meeting, transcript of public comment period, December 10, 2018.

69. FCC meeting, transcript of public comment period.

70. Zahra Ahmad, "Flint Council Votes against Paying Pipe Replacement Firm an
Additional $4.8 Million," Mlive.com, January 10, 2019, https://www.mlive.com
/news/flint/2019/01/flint-council-votes-against-paying-pipe-replacement-firm
-an-additional-48-million.html.

71. Femi Redwood, "Flint City Councilwoman Alleges Pay for Play at City Hall,"
NBC25News.com, December 13, 2018, https://nbc25news.com/news/local/flint
-city-councilwoman-alleges-pay-for-play-at-city-hall; Ahmad, "Flint City Council
Approves Additional $1.1 Million to AECOM Contract despite Pleas against It."

72. Douglas Hanks, "For Miami-Dade's Sewer Cleanup, a $91 Million Contract That Quickly Went over Budget," *Miami Herald*, March 4, 2019, https://www.miami herald.com/latest-news/article226986789.html.

73. Tribune Content Agency, "New Oakland Bridge Expected to Be Two Years Late, Millions over Budget," *Daily Republic*, January 12, 2018, https://www.dailyrepublic .com/all-dr-news/wires/new-oakland-bridge-expected-to-be-two-years-late -millions-over-budget/.

74. Yamil Berard, "Fort Worth School Trustees Blast Bond Program Supervisor," *Star-Telegram*, January 17, 2015, https://www.star-telegram.com/news/local /community/fort-worth/article6379560.html.

75. Paul Thompson, "Fairmont Austin Construction Nightmare: New Details Un-earthed in Big-Dollar Lawsuit," *Austin Business Journal*, December 12, 2018, https:// www.bizjournals.com/austin/news/2018/12/12/fairmont-austin-construction -nightmare-new-details.html; Jonah Newman, "To Build Cop Academy, Chicago Picks AECOM, Firm with Checkered Past," *Chicago Reporter*, January 24, 2019, https://www.chicagoreporter.com/to-build-cop-academy-chicago-picks-aecom -firm-with-checkered-past/.

76. Zahra Ahmad, "Sheldon Neeley Beats Karen Weaver to Become Flint's New Mayor," *FJ*, November 5, 2019, https://www.mlive.com/news/flint/2019/11/sheldon -neeley-beats-karen-weaver-to-become-flints-new-mayor.html.

77. ABC12 News Team, "Neeley: Flint Water Fund Has $20 Million Balance Instead of Looming Shortfall," ABC12.com, November 27, 2019, https://www.abc12.com content/news/Neeley-Flint-water-fund-has-20-million-balance-instead-of -looming-shortfall-565549211.html.

78. ABC12 News Team, "FDA Testing Potentially Tainted Bottled Water Passed Out in Flint," ABC12.com, January 9, 2020, https://www.abc12.com/content/news /FDA-testing-potentially-tainted-bottled-water-passed-out-in-Flint-566851211 .html.

79. Leonard N. Fleming, "Flint Excavates Lead, Iron Water Pipelines a Year ahead of Schedule," *DN*, December 4, 2018, https://www.detroitnews.com/story/news/2018 /12/04/flint-replaces-lead-pipelines-year-early/2197008002/.

80. Ninth Ward city councilmember Eva Worthy argued in a November 25, 2018, Face-book post that residents needed to voice their concerns regarding the lead service line replacement process in order to stop AECOM from generating income in Flint that had not been earned. As she put it, "We have already paid AECOM $5 million and they have not completed the 6,000 lead line replacements in the contract we signed a year ago. They have refused to use the predictive model and have dug up yards of homes likely to have copper lines. Their explanation? 'No ward left behind.' The city will have used all the funds available to dig up copper lines and leave thou-sands of lead lines in the ground. This is unacceptable! Please call your council per-son to express your opinion on the matter before the vote." See https://www .facebook.com/pg/EvaWorthingFlintCityCouncil/posts/.

81. Mitch Smith, Julie Bosman, and Monica Davey, "Flint's Water Crisis Started 5

Years Ago. It's Not Over," *New York Times*, April 25, 2019, https://www.nytimes.com
/2019/04/25/us/flint-water-crisis.html.

CONCLUSION

1. Valerie Strauss, "How the Flint Water Crisis Set Back Thousands of Students,"
Washington Post, July 3, 2019, https://www.washingtonpost.com/education/2019
/07/03/how-flint-water-crisis-set-back-thousands-students/?utm_term=.db910e
052f84.

2. Dominic Adams, "Some Flint Schools May Be among State's Lowest-Performing,
Says Interim Superintendent," *FJ*, March 15, 2018, https://www.mlive.com/news
/flint/2019/07/retired-genesee-county-employees-afraid-they-will-lose-healthcare
-benefits.html.

3. Irwin Redlener, "We Still Haven't Made Things Right in Flint," *Washington Post*,
March 7, 2018, https://www.washingtonpost.com/opinions/we-still-havent-made
-things-right-in-flint/2018/03/07/5c700692–2211–11e8-badd-7c9f29a55815_story
.html?utm_term=.5675e0921b3a; Steve Carmody, "5 Years after Flint's Crisis
Began, Is the Water Safe?," National Public Radio, April 25, 2019, https://www.npr
.org/2019/04/25/717104335/5-years-after-flints-crisis-began-is-the-water-safe.

4. Richard Rabin, "The Lead Industry and Lead Water Pipes: 'A Modest Campaign,'"
AJPH 98, no. 9 (September 2008): 1584–92.

5. W. Troesken, *The Great Lead Water Pipe Disaster* (Cambridge, MA: MIT Press,
2006).

6. Rabin, "Lead Industry and Lead Water Pipes."

7. Ronnie Levin, *Reducing Lead in Drinking Water: A Benefit Analysis*, Office of
Policy, Planning and Evaluation, USEPA, December 1986, revised Spring 1987,
EPA-230–09–86–019.

8. S.124—99th Congress: Safe Drinking Water Act Amendments of 1986, Public Law
99–339, www.GovTrack.us (1985), October 20, 2020, https://www.govtrack.us
/congress/bills/99/s124.

9. D. Chin and P. C. Karalekas Jr., *Lead Product Utilization Survey of Public Water
Supply Distribution Systems Throughout the United States*, USEPA, Boston,
Massachusetts.

10. Valerie Hoff, "2,500 Georgia Children Test Positive for Lead Exposure," 11 Alive,
April 29, 2016, https://www.11alive.com/article/news/local/2500-georgia-children
-test-positive-for-lead-exposure/159845471.

11. Lauren Cross, "Gary Gets Grant to Test Children for Lead Poisoning; City Has
Some of the Highest Rates in State," *NW Times*, July 10, 2019, https://www
.nwitimes.com/news/local/lake/gary-gets-grant-to-test-children-for-lead
-poisoning-city/article_6a6a8ce5–5f83–5209-a2dd-3d9597f817eb.html.

12. Bob Shaw, "Twin Cities–Based Water Gremlin Closes after Lead Poisoning Found
in Children," Inforum, October 28, 2019, https://www.inforum.com/news/4742835
-Twin-Cities-based-Water-Gremlin-closes-after-lead-poisoning-found-in
-children.

13. Brian McCready, "Lead-Tainted Water in Schools: Connecticut Gets Failing Grade," Across Connecticut, Patch, March 30, 2019, http://patch.com/connecticut /across-ct/lead-tainted-water-schools-connecticut-gets-failing-grade.

14. Diane Rado, "What Is the Legislature Doing about the Dangers of Lead in Schools? Not Much," *Florida Phoenix*, April 11, 2019, https://www.floridaphoenix.com/2019 /04/11/what-is-the-legislature-doing-about-the-dangers-of-lead-in-schools-not -much/; "Lead Found in Drinking Water at Area Schools," *Tallahassee Democrat*, October 25, 2016, https://www.tallahassee.com/story/news/2016/10/25/elevated -lead-levels-found-water-16-leon-schools/92732168/.

15. "All Montana Schools Should Test Their Water for Lead," *Missoulian*, July 21, 2019, https://missoulian.com/opinion/editorial/all-montana-schools-should-test-their -water-for-lead/article_dd052ed4-c4bc-55c9-bad2-e3cc83fe3e81.html.

16. Edward O'Brien, "Montana Gets Failing Grade for Lead Exposure in Public Schools," Montana Public Radio, March 22, 2019, https://www.mtpr.org/post /montana-gets-failing-grade-lead-exposure-public-schools.

17. Brenda Goodman, Andy Miller, Erica Hensley, and Elizabeth Fite, "Lax Oversight Dilutes Impact of Water Testing for Lead," Georgia Health News, June 12, 2017, http://www.georgiahealthnews.com/2017/06/lax-oversight-dilutes-impact-water -testing-lead/.

18. Susan Haigh, "Dozens of Connecticut Water Systems Violated Lead Levels," *Hartford Courant*, December 7, 2018, http://www.courant.com/news/connecticut /hc-lead-tainted-water-connecticut-20160409-story.html.

19. Molly Enking, "Flint, Newark, and the Persistent Crisis of Lead in Water," *Wired*, August 28, 2019, https://www.wired.com/story/first-flint-now-newark-the-water -crisis-is-far-from-over/.

20. D'Vera Cohn, "Investigating Washington, D.C.'s Water Quality," *Nieman Reports*, Spring 2005, https://niemanreports.org/articles/investigating-washington-d-c-s -water-quality/.

21. Marc Edwards, Simoni Triantafyllidou, and Dana Best, "Elevated Blood Lead in Young Children due to Lead-Contaminated Drinking Water: Washington, D.C., 2001–2004," *Environmental Science and Technology* 43, no. 5 (2009): 1618–23.

22. Cohn, "Investigating Washington, D.C.'s Water Quality."

23. Government of the District of Columbia, "Five Year Consolidated Plan," 78, ac- cessed December 19, 2019, https://dhcd.dc.gov/sites/default/files/dc/sites/dhcd /publication/attachments/FY2017-2021%20DHCD%20final%20Consolidated %20Plan.pdf.

24. Nancy Kaffer, "Flint's Problems Didn't Start with Water," *DFP*, February 13, 2016, https://www.freep.com/story/opinion/columnists/nancy-kaffer/2016/02/13/flint -water-foreclosure/80235900/.

25. Patrick Barnard, "ATTOM: Foreclosure Starts Fell 6% in 2018," MortgageOrb .com, January 18, 2019, https://mortgageorb.com/attom-foreclosure-starts-fell-6 -in-2018.

26. Molly Enking, "Flint, Newark, and the Persistent Crisis of Lead in Water," *Wired*,

August 28, 2019, https://www.wired.com/story/first-flint-now-newark-the-water
-crisis-is-far-from-over/.

27. Robert Bolin and Patricia A. Bolton, *Race, Religion, and Ethnicity in Disaster Recovery*, Monograph No. 42 (Boulder: University of Colorado, Institute of Behavioral
Science, 1986); Robert Bolin and L. M. Stanford, "Emergency Sheltering and Housing of Earthquake Victims: The Case of Santa Cruz County," in *The Loma Prieta,
California, Earthquake of October 17, 1989: Public Response*, ed. P. A. Bolton (Washington: US Government Printing Office, 1993): 43–50.

28. N. Jelks et al., "Mapping the Hidden Hazards: Community-Led Spatial Data Collection of Street-Level Environmental Stressors in a Degraded, Urban Watershed,"
International Journal of Environmental Research and Public Health 15, no. 4 (April
2018): 1–15, E825.

29. Thomas Craemer, "Evaluating Racial Disparities in Hurricane Katrina Relief Using
Direct Trailer Counts in New Orleans and FEMA Records," *Public Administration
Review* 70, no. 3 (May/June 2010): 367–77.

30. Walter G. Peacock and Chris Girard, "Ethnic and Racial Inequalities in Hurricane
Damage and Insurance Settlements," in *Hurricane Andrew: Ethnicity, Gender, and
the Sociology of Disasters*, ed. Walter G. Peacock, Betty H. Morrow, and Hugh Gladwin (New York: Routledge, 1997); Walter G. Peacock, Chris Girard, Betty H. Morrow, Hugh Gladwin, and Kathleen J. Tierney, "The Whittier Narrows, California
Earthquake of October 1, 1987—Social Aspects," *Earthquake Spectra* 4, no. 1 (1988):
11–23.

31. Christine L. Day, "Katrina Seven Years On: The Politics of Race and Recovery—
Notes on a Roundtable Organized for the 2012 APSA Annual Meeting," *Political
Science and Politics* 46, no. 4 (October 2013): 748–52.

32. Betty H. Morrow, "Stretching the Bonds: The Families of Andrew," in *Hurricane
Andrew: Ethnicity, Gender, and the Sociology of Disasters*, ed. Walter G. Peacock,
Betty H. Morrow, and Hugh Gladwin (New York: Routledge, 1997): 141–70; Ni-
cole Dash, Walter G. Peacock, and Betty H. Morrow, "And the Poor Get Poorer: A
Neglected Black Community," in *Hurricane Andrew*: 206–25; Elaine Enarson and
Maureen Fordham, "Lines That Divide, Ties That Bind: Race, Class, and Gender in
Women's Flood Recovery in the US and UK," *Australian Journal of Emergency Man-
agement* 15, no. 4 (January 2000): 43–53.

33. Flint Code of Ord. §§ 46–50(b), 46–51, 46–52, accessed January 29, 2019, http://
library.amlegal.com/nxt/gateway.dll/Michigan/flint_mi/cityofflintmichigan
codeofordinances?f=templates$fn=default.htm$3.0$vid=amlegal:flint_mi.

33. Alex P. Kellogg, "In Flint, a Future Built on Schools as Well as Safe Water," *Christian Science Monitor*, January 10, 2019, https://www.csmonitor.com/USA/Society
/2019/0110/In-Flint-a-future-built-on-schools-as-well-as-safe-water.

35. "A Better Future for Flint," *MSU Today*, July 19, 2019, https://msu.edu/future-for-flint/.

36. Trevor Bach, "Can Flint Be Fixed?," *US News and World Report*, February 12, 2019,
https://www.usnews.com/news/cities/articles/2019–02–12/what-will-it-take-to
-save-flint-michigan.

INDEX

Page numbers appearing in italics refer to illustrations.

DNR (Department of Natural Resources). *See* Department of Natural
Resources (DNR), Michigan
DNRE (Department of Natural Resources and Environment), Michigan,
75–76
Downtown Development Authority, 1
drinking fountains, 202, 237n30
drug use and dealing, 50, 77, 78, 105
dumping, 56–60, 83, 85, 202
Dunn Industrial Group, 70–71
Dykema, Linda, 160

Earley, Darnell, 1, 5, 81–82, 100, 149, 159
education, community, 40, 223–24n47
Edwards, Marc, 18–19, 122, 128, 129, 157–
59, 203, 220n65
emergency manager program: complaint
response and, 4, 5, 100; congressional hearings and, 149–50; corrosion
protection and, 147–48; Flint water
switches and, 81, 117; impact of, 13–16;
litigation and, 162, 165; Michigan state
and, 209, 211; removal of, 172; scholarship on, 6; water rates and, 2, 90. *See
also specific emergency managers*
emergency water sites, 92–93
Enforcement Targeting Tool Scores,
98
engineering defendants, 165–66
Engler, John: city bankrupting and, 13,
14; environmental justice and, 65, 71,
72, 73, 88–89; pollution and, 62–64,
74–75, 89
Environmental Health Outreach program, 35
environmentalism: background and
overview of, 51–52; BEN program and
(*See* Better Environment for Neighborhoods (BEN) program); opposition to
advocacy of, 62–64; trash and, 57–60
environmental justice, 204–6; Environment Justice Plan and, 75–76; Genesee

Power Station and, 65–70; Select Steel
Mill and, 70–74; urban growth and, 75
Environmental Justice Plan, 75–76
Environmental Justice Working Group,
75–76
Environmental Protection Agency
(EPA): Appeals Board of, 66, 72;
compliance testing and, 52, 129, 180;
congressional hearings and, 149; Del
Toral and, 121–22, 128–30; Flint water
recovery and, 155–56; FOIA requests
and, 19–20, 82, 154; funding from, 76,
173; Genesee Power Station and, 66,
69–70; GM plants and, 52, 108; LCR
and, 128–30; lead pipe audit by, 201; opposition to, 65; response of, 3–4, 97–98,
121–22, 128–29, 154–55; response protocols of, 118–19; safe water declaration
of, 180, 181; Select Steel Mill and, 71–75;
sluggishness of, 70; soil contamination
and, 56, 60; water crisis responsibility
of, 203, 204; water quality issues of
Flint and, 93–98
environmental racism, 70–75, 112, 205–6,
208–9
Environmental Science and Technology,
180
Environmental Working Group, 64
Evergreen Valley, 46
Evidence for Community Health Organization (ECHO), 44
Ewing, Michael, 156
Ewing, William, 86, 87
Executive Directive 2007-23, 75
Executive Orders: of Michigan, 13, 63, 76;
of United States, 72
expired bottled water, 196, 197, 207

Fahrenheit 11/9 (Moore), 136
FAST Start program. *See* pipe
replacement
Federal Housing Act of 1949, 7, 27
Fields, Kate, 178